A Tall Storey

by

Ben Jammin

authorHOUSE®

AuthorHouse™ UK Ltd.
500 Avebury Boulevard
Central Milton Keynes, MK9 2BE
www.authorhouse.co.uk
Phone: 08001974150

First published by AuthorHouse 9/6/2007

ISBN: 978-1-4343-2461-0 (sc)

Printed in the United States of America
Bloomington, Indiana

This book is printed on acid-free paper.

PROLOGUE

This is the true story of a way of life, a career and a privileged service to more than one community in more than one country, over a period from the late 1970's and into the new millennium.

It is a story for those people who stand in the street and watch those brave lads whiz past in a glowing red fire engine, in awe and admiration and it is a story about those lads and lassies with whom I have been privileged to serve over more than two decade. It is a story about people who have no idea of what really goes on inside a Fire Station, apart from the odd "Open Day" when the public are allowed in to aid charities all over the country,.

In order to protect and disguise the officers still serving those communities concerned, names and places have been changed to protect their true identities and no intentional innuendos are either meant or targeted against any individual.

This is a real story and not one invented for television. It is not just about hard work and training which we so often hear about by this dedicated band of brave men and women with whom I have for so long been associated.

It is the true story of one fire fighter's experiences of life behind the headlines, events both funny and hilarious amongst the harshness and reality of the not so glamorous moments. I hope you enjoy reading it as much as I enjoyed writing it and living it.

Ben Jammin.

Chapter One
THE WAY WE WERE

I came from military background. My father had served in the Kings Own Scottish Borderers of the Scottish Lowland Brigade during the Second World War. My older brother served as many years as a Guardsman in the first battalion of the Scots Guards after him. So it came as little surprise to my mother that I wanted to be a soldier too.

Three years after leaving school at the age of fifteen, I briefly worked as a machinist in a farm implements factory, before deciding this was not for me. The wages I was making a week were not even covering the cost of the bus to work in the next town and back each day. I was paying the factory rather than them paying me each week. My mother had adamantly stated that I was moving jobs and having spoken to the manager of a local Co-operative, she concluded that I could make twice the amount I was making at this factory if I took up an apprenticeship in Baking and Confectionery. This would mean that I had to work nights and travel twice a week to the Technical College in a major town to study for my City and Guilds qualifications.

There were very little choices in those days. My father had died when I was only two and a half years old, leaving me with one younger brother, one older brother, and four older sisters, none of whom were of working age. Mother had a very hard life to bring up seven young children on her own.

Such was the state of affairs that I often remember as a child being called in by kind neighbours to be fed a bowl of soup for dinner because my Mum just did not have enough to feed all eight of us in the family. Therefore the neighbours would take it in turn to call in the younger ones of the family to ensure we did not go hungry.

You would have thought that as the older ones grew to working age, they would have pitched in to look after the younger ones, but such was not the case. Through self interest, a head strong attitude and the need to explore the big wide world out there, at the age of sixteen, one after the other they all left home until there were but my mother, myself and my younger brother left. As my younger brother was still at school, this made any decision I had to make very difficult indeed. I was already working every day of the week, including Sundays, just to make enough for us all to live on and pay the bills. There was no other income coming into the house until my younger brother was old enough to leave school and help out.

I was in those days, however, still seriously considering joining the forces, just as my father and older brother had done before me, but I was not in a position then to let go the reins as I still had to support all three of us.

When my younger brother had left school and had been working for two years, helping to throw in his bit, I decided that this was the right moment to apply to join the forces. So with my certificates for trade tucked under my belt, a few days short of my nineteenth birthday, I joined the Lowland Brigade in the first Battalion of the Cameronions (Scottish Rifles). Originally, this regiment was born in the mid sixteen hundreds and had guarded the Covenanters of Scotland at their prayer meetings in the hills above Edinburgh having been raised solely for that purpose from volunteers amongst them.

Sadly, after three hundred faithful years of service, this regiment was seen off at Douglas Dale, Lanarkshire, a few hundred yards from where the regiment was raised in cut backs by the M. O. D. In what time I had been with the regiment, I was a trained sniper and I could have shot the whites of your eyes out at a thousand yards. All of which did me no good for, just a couple of years later, I was out of a regiment and out of a job. But I wasn't out of the military. They wouldn't let me. I had signed on for nine years and they were going to get them out of me first.

I'd spent the last two years of military service digging foxholes at midnight only to fill them in again at three in the morning in order to move to a new location where I'd dig another foxhole at five in the morning. I thought that there must be something more to life than that, and was determined not to apply for the same sort of thing again.

I was interviewed by a Brigadier from Scottish Command at Edinburgh Castle, and I thought, this time I'll be smart. You can't dig many foxholes on a boat without getting into serious difficulties. So I told the Brigadier that I wanted to apply for an inter-services transfer to the Navy if I couldn't get out of the Army.

The Brigadier advised me that the only way I could do that was if I had a trade of some description, with qualifications. This is where my Bakery and Confectionery certificate from my civilian days came into play, and I produced it in front of him. He was delighted, assuring me before he left with the certificate that there would be no problem but he would just check with Whitehall and the Ministry of Defence on my choice of posting.

So it was, some six weeks after this interview, that I was again interviewed by this same Brigadier in the barracks of my now defunct regiment. He advised me that he had received confirmation back from the M.O.D. Unfortunately it wasn't the reply we both had hoped for but at least they had replied.

There was no chance of an inter-services change to the Navy, but because of my qualifications, the recommendation was to transfer to the Army Bakery, Royal Army Ordnance Corps. A couple of months later I moved south to England and, guess what, found myself digging foxholes again at three in the morning. Only this time the foxholes were in the freezing Arctic Circle where we were watching the North Atlantic Treaty Organisation's Northern flank in Norway.

One morning, I woke up in bed and decided there and then that if I could shoot the whites of a man's eye out with deadly skill, I could systematically use every available weapon in the whole arsenal of the

Army. The only problem was that all of these have one thing in common - to kill or maim people and I didn't want to kill. I wanted to save life if I could and if the opportunity were given to me. That is what I really wanted to do above all else. The armed forces are the place where you learn the value of life. Here one minute, dead the next, which is what happens frequently in such a risky business.

The very opportunity I desired finally came along to me when I gave in my notice to leave the Army.

Six months before I was due to leave, I attended a seminar on the Police Service, Prison Service and Fire Brigade in Germany, and after watching film footage, and taking in the information provided, I decided to apply to my local Fire Brigade back home in Scotland. I flew from Germany to be interviewed and was accepted there and then. However, according to Army procedures, I still had to have an interview with the Resettlement Education Officer so that he could try and find me a job in civilian life after so many years in the forces. But his efforts were just a waste of time for me. I already had a job to go out to, one that I had found myself without the aid of the Army.

Most fire brigades are made up from a combination of two sources. The first are full time professional fire fighters, and the rest are the dedicated men and women who have full time jobs in other areas of daily life and turn out to back up the former when large numbers of personnel are required. These brave people are taught on courses run by respective area brigades and trained to work alongside the professionals until the emergency at hand is over.

Many brigades take a lot of their recruits from the armed forces, even today, people like me who have retired by right, having completed their full twelve years. To appreciate why, it helps to understand the selection process of these brigades.

Full time professional fire fighters are selected for posts after assessment in four areas: education, powers of observation, physical strength and ability to command and control individuals in a team. It is a vital part of a brigade's training to encourage individuals to exercise this ability to the full.

So with these thoughts in mind, it was off for the education test first. There were forty of us at the start and after this round, only half of us were left. Then it was on to the next round, the physical side of the testing.

This involved pairing off with another member of the twenty or so left standing and carrying each other for one hundred metres, fireman style, across the shoulder.

My partner carried me that distance wonderfully well. Then it was my turn to come all the way back with him on my shoulder. First let me just say that I carried him the whole distance. He was a young proud man, whom his mother loved, and I hated. Yet I carried him every inch of the way, all sixteen stone of him. I swear that when I started out on the journey I was six foot five inches tall but I measured only five foot six when I got to the other end. And I remained at that height throughout the rest of my career.

I also think that, by choosing him as my partner, I must have failed on my powers of observation test as the fat from his stomach banged round my ears and obliterated my sight with each pounding step that I took with this lump on my shoulder. But determined I was and, fortunately, I was selected for brigade training, the hard graft of which turned the individual into a well oiled team worker.

Chapter Two
BASIC TRAINING

Little did I realise at the beginning of this sixteen weeks' training period how much fun was involved. Maybe not all fun because, of course, as you might expect, it was hard graft. It's not intended to be a picnic. If you need to save someone's life, not to mention your own and that of your partner, you need to do things the right way. Recognising the signs of danger wherever they may happen are uppermost in your mind. So let me not take any of that away from the brigades concerned. But the lighter moments tended to ease you along your way a little, and they did.

At our station, the kitchen was over the back door. One day a fire fighter was coming in when all of a sudden a proverbial pot of water came out of the window and drenched the poor bloke from head to toe.

Not to be outdone the fire fighter raced away and grabbed a bucket of water to return the compliment on the individual concerned. Armed with this bucket of water the wet one was intent on making sure that his compatriot ended up the same way. So he ran into the station with this bucket and chased his assailant into a corner of the sub officer's bedroom on the first floor where he thought he was hiding. The proposed victim, not wishing to end up the same way as his assailant, had closed the door to the bedroom, hoping he wouldn't be discovered.

Having got this far, the one with bucket of water in hand turned the handle and burst into the room with bucket swung back ready to drown the receiver.

The only trouble was that in the room was the station officer who had been interviewing a prospective recruit in the bedroom and had closed the door for privacy. He dived for cover at the sight of the bucket, saying, "Yes, what's your problem son?" Meanwhile, the assailant had disappeared completely, leaving the one with the bucket to mumble and stutter his apologies and explain his actions to the station officer who could barely contain his laughter as he faced his would be attacker.

This was one of many incidents that happened throughout basic training.

Obviously with that much water about the occasional water fight was obligatory, and it was not unknown for an individual to be singled out as a target. In this case he'd be tied to the railings and pounded with the biggest jet a hose could supply, so hard that he could scarcely breathe and came close to drowning until the hose was turned away from him.

You really needed to have eyes in the back of your head as well as the front everywhere you went, because you never knew if it was your turn next. You had to stay ahead of the mob at all times.

On another occasion, the young secretary of the station let it slip that she was going to be married, so the whole crew, acting as one, tied her to the flag pole outside the fire station where the police found her two hours later. Not much in that you might say. However this station was on a very busy main road where she could be seen by any passers by and I neglected to mention that we had stripped her to her bra and knickers before tying her to the flag pole.

I think she said "It's a good job it's summer", or something similar when the police brought her into the station.

The chief of all fire fighters in the county are called Fire Masters, or County Fire Officers, dependent on which country you serve. They were always the be all and end all, either feared or admired by all the fire fighters, depending on which way you looked at it. And of course, they were intelligent and clever. They had to be to get to the post of being responsible for everyone in their particular county. But they still had their moments, where they joined in with their men and where, inadvertently, they were the target of humour simply by what they said or what they did to the men and women of the station. Alex was no exception.

A particularly bad fire at a chemical plant in the late seventies had seen some of Alex's men fighting the fire on a large storage tank. Suddenly they were involved in a 'BLEVE', (Boiling Liquid Evaporating Vaporising Explosion). This is where the liquid inside the tank concerned expands due to an external heat source, and when cooling is not applied quickly, an explosion occurs.

Unfortunately Alex had also been in the thick of the fire directing operations when the liquid had boiled over the top of the tank. As a result, the same fire fighters, along with the Fire Master, had to be whisked off to the local hospital for an inspection of their 'nether regions', which had become seriously overheated since in those days their leggings were made of a heavy duty plastic.

Sitting around in the waiting area of an accident and emergency ward while a fire burnt out of control, was not Alex's style. He was a man who had come up from the ranks, as had the rest of the men under his command.

After an hour had elapsed and only a few men had been treated, Alex called the duty nurse over to ask how long they were likely to be before she got through with them all. When he was told that it was likely to be a further hour and a half, Alex turned to the nurse, and said "In that case give me a handful of that grease stuff over there to rub on my privates and I shall be away up to see how the boys are getting on". Now that's a man who relates to his boys and is well worthy of his position.

On another occasion a former Fire Master of the Brigade, who had taken promotion and had moved up to the office of Her Majesty's Inspectorate of Fire Brigades, had returned to inspect a station of his old brigade. Ours. The previous watch on night shift had been out at a house fire for hours during the night and had returned to the station only minutes prior to our watch taking over for the day shift. It was eight o'clock in the morning.

Under normal circumstances everything would have been immaculate for the inspection, but there stood the inspector with a smirk on his face as he opened the fire appliance locker. He called me over as he took out the water standpipe, declaring that he could "get diseases off this thing". This guy was giving us nightmares with his attitude.

All morning we had scrubbed and cleaned. Many times we wished we could crawl out under the front door even though there was only a half inch gap above the floor of the bay. But I'm sure we could have all got under it the way this man was behaving towards us that morning.

Having made us feel so small all morning, he then disappeared to have lunch with the current Fire Master, leaving his staff car on our station forecourt whilst he did so. This gave us a good idea. In order to please him and to get him off our backs we scrubbed and polished this man's car until it was gleaming, sacrificing our lunch hour to do so.

When he returned to the station in the early afternoon, instead of being grateful as we expected, he virtually screwed us to the floor with his behaviour and stayed with us throughout the whole afternoon making our life a misery.

There is something about power that makes people disbelieve a situation, even when the evidence is there in front of their eyes. This is how it was for us when we turned up that morning and found him there. Not all Chiefs are good and I suppose you have to put it down to his job. That was the reason he was there in the first place, to ensure that our training, equipment and vehicles were up to the required standard to carry out the job we had joined up to do.

Getting a really good Fire Master is a bonus to any brigade and you certainly learn who is good or bad in a very short space of time.

Good Fire Masters think about their men all the time. Fire fighters are always coming and going through illness, retirement, or moving on to new brigades. So it was not unusual to see a good Fire Master turn up at a fellow fire fighter's retirement 'do' to see him off as it were. For at the end of the day, fire fighters are truly like one big family, all sharing the same risks, whether ordinary men and women, whether Fire Masters or County Fire Officers anywhere in the UK and throughout the world.

Chapter Three
JOINING IN

Surprisingly, sometimes staggeringly, I successfully finished the recruits' training. Not that I always played the game by the rules. Sometimes several of us hid behind the fire appliance when we should have been running round the drill yard. It was a big drill yard, all round the perimeter fence of the station. While we should have been running round five or six times, we would run round once, and dip down behind the fire appliance, out of sight of the instructor who was otherwise engaged with the crew. He was carrying out a particular drill with at the time, and we hid only to reappear on the last time round joining in with those who were stupid enough to go around the full distance. We were obviously none the worse for wear for the experience but were successful in appearing otherwise by putting on heavy breathing to deceive the instructor.

Nevertheless, when training was completed it was now time to find out where we were going and to what watch had we been assigned. For my part, I was assigned to Blue Watch.

It was never that easy to integrate right away, and by chance I had only been on the watch three weeks when I had to endure several of my colleagues being severely injured in my first real test under fire. And, I'm sorry to say, I witnessed my first fatality in the same fire.

But this is where the lighter moments bring something out in you. Maybe it helps to explain the dry sense of humour in all fire fighters the world over, which gives you a sense of balance and courage to endure what most people say they could not face.

This is one of those lighter moments.

William was a leading fire fighter, which is the next ring up the ladder from an ordinary fire fighter. William had hands like an Irishman's shovel, and was admittedly not the brightest in the land, but he was very good natured under normal circumstances. However, in the night watch round the supper table, William had been picked on by the most graciously fat bloke on the watch whom we shall call Tommy.

As was usual, barring fire calls, we had all been stood down after the work routines for the evening had finished, and the usual conversations had come up, such as who had done what during the day.

When it came William's turn, he explained he had spent the day in the next town in a brand new shopping centre just opened where he had bought a new washing machine. He thought the machine was very reasonably priced and, of course, all ears perked up at this. Not surprisingly he was asked for the name of the shop.

Unfortunately he had momentarily forgotten the name of the place and came out with, quote, "Rumbletums. You know, the place where they sell the washing machines". Poor William couldn't understand why the other four blokes were rolling about on the wooden floor in tears of laughter. Tommy couldn't resist the temptation to add the final blow the way things had been going and said, "William, you're stupid".

We all retired to the rest room where we watched the evening news on TV, and not another word was spoken after Tommy's comment for around half an hour. All of a sudden, William turned to Tommy, intent on getting his own back, bearing in mind the latter's eighteen stone hulk. He intended to comment on the fact that the latter was grossly overweight for his size. However, while meaning to say that Tommy was obese, instead put his foot in it even more by remarking "Tommy, you are *obsessed*". Once more the tears of laughter could be heard from the other four on watch that night, and to this day I still don't know if William ever knew why?

William was a grand storyteller, who told it the way it was. On occasions we all sat round the supper table on watch discussing each other's days as a boy, and we would take it in turn to relate how and where we each grew up, what we did, and how we would spend our time. I loved to tell how I would go up on my own and sit amongst the bracken and fern of a high vantage point in the nearby woods and look out over the local river, to the sight of the famous three bridges of Scotland. These could clearly be seen from my position of height.

Somewhere along the line, the subject of friends of younger days was introduced into the conversation and we talked about the games we played to pass the time.

Everyone in turn explained what friends they had and how each would go down to the local woods, build swings on trees, or a cave from branches and pretend this was the club house, like you did in those days. Not having much money then, everyone was in the same boat.

That was until it came to William's turn. He explained that, unlike all the rest of us, his best friend was a girl and not a boy. Taken aback as we were by this statement, we were even more interested in finding out what their favourite game was. Until the arrival in the mid sixties of the diesel locomotive, all trains were steam engines. William lived in a village where the local railway station was virtually on these two young friends' doorstep, and the story he told had us splitting our sides with laughter.

It seems that William and his little girl friend would wait until a steam train arrived in the station and, standing on one platform, have a competition to see if they could pee over the top of the carriages of the train and land on the platform on the other side.

William was saying, that try as he might, he could never get over the top of the train, but this little girl always did it. Apparently she had an enormous pee, and every time she made it to the other platform, so he could never compete with her. You can just imagine the roars of laughter this brought about around the table, but as I say, William always told it the way it was.

If there is one thing I am trying to get across here, it is just how close a team you all become when you work together day in day out. And yet each and every one of us still had our individual lives with wives and children outside this somewhat artificial atmosphere.

Chapter Four
RUNNING WITH THE TIMES

There were many not so good times of course. One of the most horrendous, and yet endearing moments had to be when a petrol tanker slid off the road because of weather conditions at the time.

We had a battle just to get to that one as it was early March and we were some three miles away in a heavy snowstorm with strong winds, which created a white out for the driver who could see no more than a few yards ahead.

Reaching the scene of the accident, we proceeded to cover the trailer of the unit, in case of fire outbreak, by placing a fire hose linked to the fire appliance with a fire fighter on the end of the jet.

Places on the appliance, as well as jobs, change from day to day to make sure that all who serve are given the opportunity to be seen to share the work load and the dangers. Unfortunately for me, it was my turn that day to fight the fires.

Now to do the job correctly it was necessary to put one foot forward and lean into the jet, to counteract the reaction when a branch at the end is opened and the hose is already charged with water. This is called jet reaction, and we had to take this precaution just in case we had to open the branch quickly.

In the brigades they have a statue in bronze which they sell for charity, and which depicts a fire fighter doing exactly this.

The snow was drifting even where we stood because of the temperature and the wind, and after two hours of being on the end of this hose, I looked exactly like the statue of the fire fighter.

The wind was blowing directly into my face, and the accompanying snow had built up on the front of me to a depth, I reckon, of two feet on the upper part of my body as well as on my leading leg where I had stepped forward and leaned into the hose.

When the rest of the crew came to relieve me from the post, they had to peel me off the hose, for I was frozen. I couldn't move my knees or arms, and was still stuck in the position where I had gripped the jet with one hand under and one hand over. By peeling my fingers off the jet one at a time they eventually managed to free me from the hose.

I was later to discover that while I had been outside literally getting frozen stiff, the rest of the crew had been sitting in the cab of the appliance with the driver of the tanker, and having a cup of tea from the driver's flask that he was carrying.

As if that wasn't enough, the driver of the petrol tanker had previously stopped on his route and bought six eggs for later use, and put them on the dashboard of his truck at the time of the accident. Such was the gentle descent of the tanker sliding down the road that those six eggs were still sitting there totally undamaged.

The only one who seemed to be damaged that day was myself. I'm sure that if the remainder of the crew had pulled me back after releasing my fingers from the jet of the hose, my foremost leg would have been stuck in the bent position, with my arms sticking out front. If I had hit the ground at any point I'm sure I would have shattered.

The subject of Fire Masters and County Fire Officers, reminds me of another incident we were called to involving a colleague called James. On another night shift, we attended a road traffic accident involving a cattle float with two layers of cows on the trailer. The poor cows were trapped in distressing conditions, so speed of release was of the essence.

When the float had left the road, it had ploughed into a plantation of young fir trees and ended half way up an embankment. As the motor cab of the float began to topple sideways after being unbalanced, the front of the cab had taken the roof off by striking a young fir tree. The tree had acted like a catapult, bending back and hurtling the driver right out of his cab and onto the road. However, we still had a co-driver to find when we turned up at the scene.

Eventually, after searching the surrounding area for the co-driver, we heard faint cries from the underside of the cattle float trailer. On investigating we found that the co-driver had been hurtled from the cab by this same tree and was in the very precarious position of being trapped beneath the trailer at the foot of the embankment. The trailer side was resting against his nose as he lay on his back, trapped by another young fir.

What had happened was that when the cattle float left the road, it ended up in a straight line, half way up the embankment, with the wheels precariously perched and the trailer on its side at the foot of the embankment, leaning on the outside edges of the tyres on one side.

Five feet along the trailer to the back of the cab this young lad lay pinned underneath.

Under normal circumstances, we would have stabilized the unit at the wheels by hammering in chisels or such like to prevent the unit from sliding further down the embankment. We would then have called in a heavy crane to put the unit on its wheels again. But here we were with this young lad flat on his back and the unit pinned up against his nose five feet along the trailer with two decks of cows trapped inside and under exceptional stress.

If that wasn't enough, because the unit was at an angle, the cows inside were all piled on top of each other and were trying hard to get upright. As they were attempting to push their way with their hooves and legs as they struggled to upright themselves, some of their hooves pushed through the side slats of the trailer. This made it very dangerous indeed for the co-driver, as some of their hooves were only centimetres away from his head and any point of contact with him could have killed him instantly. We really were in trouble with this one.

As a team, we all got round and discussed the best method to release the co-driver, but the more we discussed it, the more apparent it became that no normal means of rescue would do as the slightest movement of the trailer would definitely crush him to death.

Foremost in our minds was the fact that these cattle were still thrashing about inside the trailer and could have moved it themselves, bringing down the embankment on top of him.

The distress of the cattle was also of great concern to us, because if we could not lessen the weight problem and movement in the trailer, the chances were that this young lad was not going to see another day before we got to him.

After a few minutes discussion it was decided that, because of the precarious position of this young man, we should hammer chisels into the ground along the foot of the trailer to prevent any further movement down the bank. As we were starting to do this, the heavy lifting crane arrived on the scene.

Having discovered that because of rain previous to the incident, the ground was too soft for the chisels to hold all of the weight of the trailer and its load, it was agreed that the chisels were insufficient for the job. We decided to dig a channel to the co-driver wide enough for one man, a paramedic, to crawl to him and pass him oxygen and a blanket as he was already showing the first signs of shock. We would then dig a second channel beneath the trailer, far enough away from the young man to put in place a heavy duty canvas belt, called a 'strop', and position the crane over the top of the embankment. If the trailer did start to slide, the strop would prevent it from doing so by holding it in position. At least that was the idea.

To be able to do this we had to pass the strop under the trailer's body, attach it to the crane hook and have the crane take up the tension.

That was our first, and luckily, only mistake. As the crane took up the strain on the strop to prevent any further movement, it did so with a twist which started to move the trailer the one way we did not want it to move - down the bank.

I guess God, and only God, was with that co-driver that night as the young man saw it coming and managed to turn his head to one side. As it was, the trailer moved only a fraction of an inch and settled again.

We as a team then decided that the only safe way to release him was to go into the young forest, about ten metres back, and dig under a perimeter fence, with whatever it took until we could come up in a narrow channel beneath him and release him that way. We could see by the use of a light that he was trapped by another young fir tree across his feet, which with the weight of the trailer was immoveable. We would have to cut it away when we could reach him in safety.

It took us all night long and into midday the next morning, working in relays, to dig this channel and release the young man. This is where James came into the picture.

Often, in a long drawn out event, messages must be passed back to control by radio, to keep them updated on the current situation. As it happened, James was out and about returning from an official function when he heard the report on the brigade radio in his car.

Being the Chief, he decided to go along and offer his men moral support. He was like that was our James. He duly arrived on the scene, and asked for an up to the minute report on the situation from his sub officer in charge, to see if he needed to take control of the rescue.

He agreed that our course of action was the young co-driver's best chance of survival and went along with us but James wanted to see the young man's situation for himself. With the aid of a torch, he crawled as far as he could into the narrow hole we had made earlier for the paramedic's access to the young man.

Surveying the situation, he spoke to the young man and ascertained that he was as comfortable as could be expected under the circumstances, re-assuring him as he went. He crawled out again and spoke to us saying that he was satisfied but saying at the same time, "I swear that ...." and left it at that.

All through the night the Fire Master stayed with us, giving us encouragement, and often helping with the digging to reach the young man. Such was his nature, and I could not fault him for that. Eventually we reached the lad and with a saw cut away the one long slender tree that had kept him there all night long.

We pulled him free, and he was able to be placed on a ambulance trolley. He was still conscious but suffering from hypothermia and shock. James finished his sentence from earlier on. "...it is," he said, "He's one of mine". He then went over to the young man on the trolley and said, "Hurry up and get better son. Your Fire Master wants a word with you."

It appears that the young man in question had been earning an extra few bob by moonlighting, and what the Fire Master had said to him would have been insignificant after his ordeal.

Chapter Five
THE OLD, THE NEW AND WEAR TOO

In the earlier years fire appliances, although well stocked for any emergency, were what one could only describe as leaving a lot to be desired. There were none of your fully automatic flashy machines then.

Instead we had this old Bedford fire appliance, a four cylinder petrol thing, with a door in the middle where the crew climbed in behind the driver. When you turned out to a call or an emergency, until it had warmed up, it used to putt and bang all the way down the long road to the town in question. I'm sure it must have been a re-assuring sound to the public, however, for if you were waiting for us, you always knew we were on the way. You could hear us for miles before you saw us. Such was the state and power of these machines that when you were called out, if the appliance had to climb up a hill, you could get out, leave the driver to drive up, and beat him to the top, by walking.

So it was a blessed relief when, about a year later, these were replaced on a rolling programme and each station was upgraded with motors with doors on each side for the crew to get in and out, and a separate compartment to get dressed in the back on the way to emergencies.

In the earlier years, radio communication was also, shall we say, lacking that certain touch of clear understanding, and often messages back to control would have to be repeated several times to be understood clearly.

On one such memorable occasion, we had been called to a house fire at the other end of town, and we were having trouble with a crowd of young lads. This was the rough end of the town, where if you didn't live there, you had no right to be there. You know the sort of place.

Anyway, as a precaution, my sub officer had asked me to put back a radio message to control asking for the attendance of the police to ensure our safety while we got on with fighting the fire.

This I duly did, and we were busy getting inside the premises to bring the fire under control, when about ten minutes later, as I was at the appliance locker to get a piece of equipment, a very soft spoken voice behind me said, "Where are they then, my son?"

When I turned around, there in front of me was this man in full cassock, complete with bible, saying he had been telephoned by our control to attend the incident and had cycled all the way from his church house in another part of the town.

Our control had got my call so badly wrong, that instead of sending us the police, they had sent us a priest to administer the last rites to some poor soul who wasn't inside anyway.

On another occasion, we had been called to an electrical sub station in another part of town, where it had become apparent that the fire inside had been deliberately set by young children. There were burnt out newspapers everywhere and two young lads kept pulling on our tunics and saying, "It was two boys who live over on the houses opposite Mr. We saw them."

As you can imagine, this is a very dangerous practice which every fire fighter and electricity worker will discourage, not just for the fire itself, but the very fact that simply by being in such a place, they could have been killed by electric shock.

It was decided to send for the police and frighten the living daylights out of them, in order help them understand that such a practice was so dangerous. So we put back a message to control over the radio to that effect and ten minutes later our policeman, whom we named 'PC Plod', turned up.

Now he was a big man, well over six feet tall, and this sub station had been built with a brick wall around its perimeter and a tall gate with a full frame so that it could be securely locked. Yet, for some unknown reason, the brick wall was only three feet high, allowing the kids to get into the place quite easily.

Now, being such a practiced policeman like he was, PC Plod got down to the business of interviewing the two children we had with us. Having completed this side of the matter, he then decided to see for himself how the children had managed to set the place alight and also to see all the newspaper they had left lying around.

After being advised of the dangers involved by the leading fire fighter in charge, and being told that the power to the sub station was still being provided with current, our PC followed the leading fire fighter under the frame of the doorway. Our leading fire fighter stopped and warned him how low the top of the frame was.

PC Plod removed his uniform cap from his head because of his height and followed the leading fire fighter in. The leading fire fighter ducked before going in but our PC did not, and the door frame caught him smack bang in the middle of his forehead, so hard that the force shook the frame to its roots.

Had it been anyone else but fire fighters they would have rushed over to make sure the man was OK. But we were fire fighters, and as it was we were round the back of the fire appliance, curled up in fits of laughter at the stupidity of this man.

I'm sure we must have upset him by our outburst of laughter, but determined not to be put off by it, he followed our leading fire fighter into the sub station where he walked around for the next few minutes holding a very sore head. He then surveyed the evidence before following the fire fighter out again under the frame of the gate.

He remembered to remove his uniform cap again but as our officer ducked on the way out, PC Plod forgot once more and, with the almightiest bang, caught the door frame smack in the middle of the forehead again. What a bang.

That was too much. The whole fire fighting crew were rolling about on the inside of the cab and on the ground at the plain unintentional stupidity of the man.

Somehow he then managed to half fumble, half stagger away to warn those kids of the dangers.

The strange thing is we never again came across this man locally. Maybe he got a transfer to another town with bigger gates.

At another time, we were called to a fire in a block of Orlit houses in the next village. Orlit houses were made of metal and this one was two storeys, semi-detached – four houses in all.

On arrival we were confronted with the downstairs house on the left of the block well alight. The fire had also gone upstairs to the house above through the living room ceiling, which had collapsed. It had set that house alight, gone into the rafters of the roof, and due to the absence of a good fire dividing wall in these older houses, had passed over to the right hand top house. This was also well alight and because the floor of this house had burned through, it was dropping burning timber into the house below that. All four houses were now on fire.

There was only one crew of five men, and one of us had to stay with the fire appliance to pump the water in. Twenty minutes later we were to get our second appliance from the next main town.

However, undaunted, we set about putting the fire out, starting first on the top left hand house, working our way down to the bottom left hand, then across to the right hand side of the building, with the other crew who had by this time arrived to help out.

It was decided that the second crew should concentrate on the left hand side of the building whilst we in turn would deal with the right hand houses, which by this time were well alight.

We needed to switch off the power which supplied the electricity to the houses from the street because of the danger of electrocution from lots of water on metal houses. So we called in the local electricity supplier who sent a man and van to the scene to disconnect the supply.

When he arrived, my sub officer in charge asked me to escort him and show him where the main supply was so that he could disconnect the fuses to the supply. However, by this time lots of water had been flung onto the houses by the hoses we were using. The fuse box was in a cupboard in the bottom right hand house, which meant us wading through water a foot deep to get to it.

While I held the electrician's tool bag, he got inside the cupboard and opened up the box containing the fuses. These were not normal fuses, but ones where the power comes in from the street at four hundred volts and the eighty ampere fuse reduces the power to two hundred and forty volts for domestic use.

After reminding the man of the dangers of water and electricity, I asked him to put on some special gloves I had with me which are tested up to three thousand, three hundred volts, before removing the fuses from the box. But I was dismissed by this man, saying, "I have personally done this often with bare hands and I know what I am doing."

I have never seen such a pop idol in my life as large as this man. As he reached over to pull the first fuse, there was an almighty bang and his hair shot up on end like a present day pop star, his teeth chattering together with the sheer strain of having to let go the fuse.

I had to rescue him as well as putting out the fire with the rest of the crew, and he was in a right state for half an hour. Thankfully he did come around to his normal self after that period, and went back in with the gloves on this time and pulled the other fuses, assuring us before he disappeared in his van that the place had no electricity and was safe now. So we set to work on putting out the fires which by this time had reached both houses on the right hand side which were well alight.

To achieve this we needed to get inside the top house of the two, via a row of steps leading from the garden up to the door. On opening the door and climbing the concrete stairs to a hallway, which led to all the rooms, the three of us were confronted with a house which was very much ablaze, and with no rooms off the hallway left untouched by the fire.

The smallest lad of the watch was at the front of the hose. Tommy, the big eighteen stone one, was in the middle and I was at the back of them hauling the hose up the concrete stairs. We made our way forward to the end of the hallway fighting the fire and pushing it back in very hot conditions, where your ears tingle and burn from the steam. In those days we had no fancy flash hoods as they now have, and as I had in the latter stages of my career. Just bare hands and no gloves then.

All of a sudden the first roof tiles began to fall from the apex roof and land on our fire helmets with a heavy crashing sound. It must have done something to the little lad in front. In these conditions you could only see as far as the fire in front lit up the place. And it was very smoky. One minute I was at the rear of the other two, and the next moment when a flash of fire lit up the hallway, there was no sign of the other two and I was left in front as more tiles came crashing down from the roof. When I looked round, there were the other two now behind me.

Now we all know that fire fighters are heroes. Everybody will tell you that. But with all this noise and debris falling, a dead hero is not much good to anyone, a fact that my compatriots had realised before I had. So as soon as it smoked over again, I was off like a shot and miraculously found myself at the back of the three again. However, I reckon that this situation must have lasted for all of thirty seconds, for when the fire lit the hallway again there was no sign of the others in front of me, and there they were behind me once more.

It was at this point that our sub officer and leading fire fighter from the crew joined us with another hose to help fight the fire and between us we beat it out. It took us all night to do that. Black, covered in debris and utterly exhausted but feeling very pleased with ourselves, all five of us made our way to the concrete stairway leading down to the doorway.

We were in single file from the top of the stairs to the bottom, on steps freely running with water like a river with the leading fire fighter in front, the sub officer and the little lad of the watch behind, and myself and Tommy, the eighteen stone one, bringing up the rear.

That was until Tommy slipped on the edge of a wet step, stumbled, and fell over, pushing into the back of me with the force of all of his weight. I in turn was bowled over, and crashed into the back of the little lad. He crashed into the back of the sub officer, who crashed into the back of the leading fire fighter, who had reached the open door by now and was last seen hurtling out. A few moments later, after regaining our composure and our feet, and wondering what had happened in this skittle effect, we found the leading fire fighter outside in the garden where he had landed after being thrown out of the doorway. He had crashed over the wall surrounding the stairway and was lying flat on his back moaning. Luckily he wasn't hurt by the fall.

Before we could finally leave the scene in the morning, we had to ensure that the fire was completely out. To this end our sub officer decided to climb to the top of the left hand house, which by this time had lost all of its roof timbers leaving only the metal frame of the building and a one foot wide metal ledge where the beams had formerly rested. Like the rest, this was very wet and slippery. Getting the metal ladder off the appliance once more, we set it up so he could survey the damage from the top.

When he eventually got to the top he was shocked at the total devastation and called the leading fire fighter up to have a look. Not to be left out, the rest of the crew didn't wait for an invite, but climbed the metal ladder to the top of this ledge and joined in anyway.

Now this block had forced air heating from ducts throughout each house, and as we looked down on the devastation on the left hand side of the block, we could plainly see, in a separate compartment, the fuse box intended to heat it. That was when we realised that the electrician from the previous night had turned off the electricity to the right hand side of the block, but not the left hand side where we were now standing on a wet narrow ledge made of metal.

All of a sudden there was a buzzing noise and out of this compartment that we were looking down on from the top of a very wet metal ledge, came a blue flame like a small rocket, travelling along the thick cable and straight towards the metal frame of the building just below where we were standing.

We looked at each other in horror, suddenly realising what would happen if the jet of flame reached the frame we were standing on. Without a word being said, we all as one made a bolt for the ladder. It was like the Muppet Show, with five men fighting to get on a ladder made for two at the most and where somehow the last on managed to be first to the ground. We were climbing on shoulders, backs, and anything to hand in order to get off that metal ladder. Luckily, we all made it.

Chapter Six
FIRES ARE A SERIOUS BUSINESS – REALLY?

Basically all brigades throughout the country tend to stick to the same training and methods. But individual brigades have slight variations on how they tackle certain types of fires. Take chimney fires. Thankfully these are gradually lessening now, with the introduction of central heating in most homes. But open fires were in their heyday in the early eighties, to which this period of the book relates.

Our specific method of tackling chimney fires in those days would always be from the top. We would climb up to the roof by ladder, or out from a skylight, and head for the chimney. We would then tackle the fire by cracking the hose reel open at the branch and running water round the pot on top and down inside.

On one such occasion we had been called to a chimney on fire in a village on our ground. As was usual, we always worked in pairs on the chimney for safety reasons. My partner had gone ahead on the ladder we had set up and I was preparing to follow him up.

The usual method would have been to take a second ladder which we were carrying on the appliance, and throw it over the ridge of the roof to reach the chimney in question.

On looking at the property from the ground, the pitch of the roof looked a lot less steep than the usual one. So it was decided that a perfectly acceptable method would be to use only one ladder to reach the start of the roof and use the corner ridge of the roof to climb up and shimmy across the ridge to the chimney itself.

My partner climbed the ladder first, with the hose reel tied in a loop over his shoulder facing to the rear of him. He reached the top of the ladder and climbed the ridge tiles to the top of the ridge as we had planned. Then it was my turn to follow him up.

Unfortunately, I didn't have such an easy ride, as my partner on reaching the top had caught the hose reel branch on the ladder. This in turn had caused it to open, allowing water to run all over the roof as I tried to scramble across the roof to the corner ridge and follow him up. Much worse was to happen as I soon discovered. We had found the only tiles in the village coated in tiny loose sandstone pebbles, which when wet broke away from the tile.

As I climbed towards the corner ridge, I felt my feet start to go and by the time I neared the ridge, my legs were going a hundred to the dozen, trying to stay upright, as these little balls of sand rolled out from under my feet. I failed miserably, and with a crash hit the roof face down and started to slide towards the edge of the roof in free fall.

There are only certain things you can do to delay or prevent this, like pray for example which I did. I also spread-eagled my arms and legs to slow down the slipping process, while screaming to the rest of the crew, "Get the b***** roof ladder across the ridge now!", which they did with the speed of light. Thanks to their quick action, I was able to grab it and pull myself onto it to safety. Otherwise that day might have had a very different outcome.

In those days, and until very recently, breathing apparatus was a heavy cumbersome affair weighing an extra four stones on top of your fire gear. So the two who were going to fight the fires were always positioned on the outside of the cab on either side in order to be ready to tackle the fire as soon as we got there.

One day we were called to a house fire and on approaching the address smoke was clearly seen coming from the property, so we both donned our masks and turned on the air going ready to go. As soon as the motor stopped, I jumped out of my side of the cab and grabbing the hose reel from the side locker, I waited for my partner to join me from the other side of the cab, nearest the pavement.

After what seemed like an eternity of waiting there was still no sign of him. So I dropped the hose reel and ran round to find out where he had gone. I hadn't expected the sight which met my eyes when I came round the corner. There was my partner by the side of the road with his leg stuck down a rainwater drain. The poor man's shin bone was scraped down the front where he had gone in and he was firmly wedged by his Wellington.

It seems that he had jumped out of the motor as I had done at the side of the road, but on his side, the step was level with a drain and someone had kindly removed the drain cover.

Another time we turned out to attend a chimney fire, and as the appliance drew up at this lady's door,

I jumped off onto the pavement. I was dressed in fire helmet, yellow leggings, black tunic and Wellingtons. The woman ran up to me and said " Thank goodness, are you the fire brigade?" I looked at her and quick as a flash replied, "No madam, I'm the cheer leader. The fire brigade will be along shortly." She promptly fainted on me. Too much excitement no doubt.

We had occasion to attend a chimney fire in a multiple block of flats with its many chimneys. When we reached the chimney concerned, to make sure that we knew we had the right chimney, the technique was to throw a few pebbles down from the top of the pot. Then the man at the fireside would shout up if he could hear or see these pebbles coming down to confirm that we had the correct one. When we heard the confirming shout from the man at the fireside, we set to work on opening the branch and spraying.

The house affected was downstairs on the ground level and often in these multiple flats they used borrowed vents, where a chimney from one or several houses, is tapped into a single chimney stack at an angle. Although this particular chimney head was hot on arrival, we had been pouring water down it in spurts for about twenty minutes and there was still no sign of water or droplets at the hearth as we might have expected.

We were perplexed by this and we asked the gentleman concerned if he had another fireplace in the house that the water might be going to. He replied that he did, in the bedroom, but he said it was never used and had been boarded in. We immediately dispatched the man at the hearth to investigate. There was no sign of water there either, so we sent the same man outside the premises to check the walls for signs of water. It was possible that the cement inside the chimney had crumbled through the years, allowing the water to run into the wall. However, there was still no sign of the water.

It was at this point, while our man was still outside the premises, that a gentleman from the right hand flat came out saying, "Excuse me, are you nearly finished? It's just that I have a little water coming out of my fireplace and I can no longer handle it. I've been trying to contain it for the last twenty minutes."

We raced into the flat of this gentleman and we found his brand new carpet, and all his new furniture floating about in eighteen inches of water from our hose reel. Sorry sir, we didn't mean it.

Finally there was the time of our first encounter with the very first policewoman in our area. She was a shining example to the force, straight out of police training college and on her first posting to a station.

A little after six in the evening we received a 'persons reported' call. This is where a person or persons are reported to be still inside the affected building and where the building was seen to be issuing smoke from it.

That night it was myself and the sub officer who went in to tackle the fire and try to find the missing man after neighbours at the doorway of the house told us that he was still in there. When we got inside it soon became apparent that the fire had been burning for some time undetected, before it was reported by passers-by.

However we set about putting the fire out and searching each room in turn to try to locate the individual. He wasn't in either of the two bedrooms, which only left the living room and the kitchen. We had no luck in the living room so we made our way to the kitchen. When we opened the door, there in front of us was this very unfortunate gentleman, and it immediately became apparent that we were far too late to save him. He was long since gone.

We put out the fire in the house and retired to the open air, leaving the poor unfortunate gentleman in the position we found him. This is the procedure when such occurrences happen, as the police will wish to take photographs of the scene for a coroner's inquest at a later date.

When we got outside, we took our breathing apparatus sets off and awaited the arrival of the police to investigate the scene. We had put in a request over the brigade radio and there within a few moments our very first woman police constable arrived.

Now she was a very thorough policewoman, and it was apparent that her male colleague, who had been assigned to look after her in her first few weeks, did not want to make out the report on this one. So he dutifully sent her over to take details. As she approached the sub officer and myself, she took her notebook in hand obviously meaning business.

She asked who was responsible for finding the dead person, to which we replied that we were. Having ascertained that fact, she then asked our names, date of birth, the ranks we held in the brigade and what each was doing when the body was found, noting down every word of our reply in her notebook. Then she asked the fatal question, "Was he dead when you found him?"

At this point I'm afraid I must own up to being the culprit. Noting that she was writing down every word verbatim, I replied, "No officer, he was alive and I had him in my arms." She then asked me if he said anything before he died, to which I replied, " Yes, officer", he said "That's funny, I've never done this before." and I promise you because she was still new to the game she wrote down every single word in her notebook.

Then she asked, "What happened next?" My sub officer replied, "We opened up the Hearth Tool Kit," (which is a box of tools for doing odd jobs) "and got out the tools. We were about to carry out open heart surgery when you turned up." She wrote down every word as gospel, and she really didn't realise that we were getting at her.

Concluding the interview with us, she returned to her male colleague. He asked to see her notebook and the statement we'd made to make sure she had it right. As he started to read through the statement, he burst into tears of laughter and it was then that it finally hit her. She had been feeling proud as punch up to that point after concluding her first interview, and she came after us with a vengeance. She was a big woman, but we couldn't stop ourselves laughing, and to her credit she did very soon see the funny side of being so well duped and joined in the laughter.

Chapter Seven

WHERE YOU'RE AT IS
WHERE YOU ARE.

Not all fire stations are built in the middle of major towns or cities. Often these stations are positioned where they can cover major risks in the area concerned, and these may well be large areas of rural population and land.

The one thing all fire fighters have in common, whichever brigade you join, is that you have little or no say as to which station you will be posted. You might wait for several years of your career before getting to a station of your choice. It is stated policy that each individual has the opportunity of promotion to a particular post through passing preset examinations followed by interviews.

However, the truth is that, even having passed these exams, you have a much greater chance of promotion if you join a big city brigade with lots of stations, than a rural brigade with only a handful of stations. A rural brigade will only have a limited amount of places in the rank structure to accommodate your ambitions. Anyway, not everyone wants to be a whiz kid, where the further up the ranks you go, the further away you are from the fire appliance itself. Through the diversity of different jobs in a particular brigade you can climb what is commonly called the ladder of success.

If you're posted to a city brigade, you might find yourself with a couple of miles of high flats to look after, and one or two industrial sites thrown in. Whereas in the rural stations, you're more likely to be looking after an area of many square miles, incorporating industrial sites considered high or low risk dependent on the nature of what is made in the building and how flammable it might be if involved in fire. Alternatively, you could have a major town with many villages surrounding it, and rural land and forests, mountains and hills. It all depends what station you are serving.

In my case, in the early years, I was on a rural station containing hills and mountains in its ground with all the sorts of risks that this implies. All it took was a member of the public to either telephone the emergency number or walk into the station and say they had seen a fire on the hills, and off we would go. We headed for a predetermined car park where the driver of the appliance would remain with the machine. The rest of the crew disembarked, got off the beaters for grass and with leggings, Wellington boots, tunics and helmet on, start to climb to a height sometimes in excess of two thousand feet.

Usually the officer in charge had a hand held radio, which in theory he used to relay messages back to the driver of the appliance below. In theory, the driver could then relay the present situation on to our control. But, let's face it, in those days these hand held radios were absolutely useless at a range of more than two hundred and fifty metres and the driver would hear nothing from the crew for many hours until they re-emerged from the hills and could pass on the message themselves.

It was not unusual to get off an appliance at nine o'clock in the morning, climb the mountain and not get back down to the appliance until dusk. We often had to climb to a height of two thousand feet straight up, or go back two or three hills in the range to climb to the same height to reach the fire.

Very often, by the time the crew reached the fire the gorse bushes or grass concerned would have burned itself out and all you were left to do was to check that it was fully out and return to the appliance from whence you came. But that's all part of being a fire fighter.

However, if you were lucky enough to be the driver that day, it was a good day out as there was nothing else you could do but rest in the back of the motor in case control contacted you for an updated report. And all you really could say then was that the crew were still on the hill and that the appliance would be tied up for some time so they could get another crew from a nearby town or village to cover the station ground, as is the practice.

There was an absolute classic incident concerning a part-time station not too far away from our own which was also rural.

Manned by part-time men doing other day jobs, this crew had been dispatched to a road accident way up on a mountain road. It was many miles from their own station but they were the nearest appliance to the accident.

That day, five men with a sub officer in charge, manned the fire engine. En route they made multiple calls for assistance to our control.

Communications were very poor in a mountainous region in those days, and control wanted to ascertain how far away from the incident the appliance was at that time. So the question came over the radio from our control, "Appliance, what is your present position?" The response from the sub officer should have been "Appliance number nnn, we are n number of miles away, or our present position is . . ."

Instead, what came back from this real country Scotsman was "We're making haste, Hen. We're making haste". It had us rolling about in fits of laughter in the back of our appliance. Only a country boy could get away with that, and he did get away with it for I think no one in the brigade could stop laughing long enough to reprimand him.

Fire fighters can be a funny lot though, in their own way. Not all days are spent running in and out of the station on fire calls or emergencies. On quiet days, where no calls are received, there is nothing to do apart from the daily routine of training and cleaning equipment in readiness. If this goes on for several days and nights, something happens to half a dozen men on watch cooped up in a station in constant readiness.

Nerves begin to fray, and where normally passive men and women of a mild nature dwell, their attitude to each other begins to change and signs of personality clashes emerge. Suffice it to say that on such days, when one can observe the general public coming and going on their daily business, a fire station can feel like

Pentonville Prison. This is especially true on days of public holidays where normally you are allowed to stand down from the usual work routines. God help it if this situation lasted more than two or three days or nights, for then they were really at each other's throats.

To relieve such boredom and tension it was not unknown for little games to be invented where fire fighters would challenge each other. For example, from a standing position, one would jump up and see how high he or she could reach with their hand on the wall above muster bay where all the fire kit was kept on pegs in readiness, next to the fire appliance.

Naturally, the next in line, not wishing to be outdone, would just have to get higher, and so it went on. To state the obvious, the taller you were, the more advantage you had, so the smaller ones at that game used to cheat a little by taking two or three steps back when no one was looking and then jump, recording their particular mark on the wall with a pencil.

This would go on until it got to such a state that someone would end up flat on their backs on the tiles of the muster bay, really hurting themselves in many cases. So then we would have to devise another game a little less harmful.

One of my particular crew had a trick of being able to raise an old fashioned weight of fifty-six pounds from ground level to his shoulder, and straight up above his head lifting it through the ring on the top by using his little finger only.

Nobody else on the crew could do it, no matter how hard we tried, and in many cases, we were in danger of getting a very sore head even attempting it. We all knew that there must a knack to doing this, but as yet remained baffled as to how it was done. So we never stopped pestering him to repeat it, which he was only too glad to do. But he wouldn't give us an insight as to how he did it.

On station we had a cook and the kitchen was her domain, apart from the times where we would make a quick cup of tea in the morning break and she would work around us, pottering away at that day's dinner. One day, the cook had been in for half an hour and had been cleaning up the kitchen in preparation to cook, when we, as one, wanted this member of the crew to show us his trick with the weight.

The weight, however, was downstairs in a cupboard. Even though he didn't have it with him, he wanted to show off his masculinity just the same. While we didn't have the weight, we did have a new fifty-six pound bag of potatoes we had just bought from the fruiterer for the cook to use and this was in a paper sack. Our colleague said the trick was so easy that he could do it with the sack. We, of course, being disbelievers, would have none of this and told him to prove it. So he promptly reached down, hooked up the sack with his little finger and, without a moment's hesitation, picked it up and raised it above his head in the same manner. Everybody stared in total disbelief as he held this sack of potatoes high above his head. And that was where it all started to go wrong.

The cook was over the kitchen sink washing up and the man with the sack above his head was in the middle of the kitchen, when suddenly and without warning came this tearing noise. The sack ripped from end to end and all the potatoes and all the dirt came flying out in one great burst of dust landing all over our cook's beautifully clean kitchen.

We all looked at each other for a split second, and knowing the temper of our cook, without a uttering a word we raced out of the kitchen. She flew into an understandable rage when she realised what had happened.

We bolted out of the kitchen and through the dining area. As luck would have it, we were all trying to get out of this one dividing door to the recreational room together and the perpetrator ended up last in the queue. It clearly wasn't possible for us all to get through together but we were trying. There was also a serving hatch connecting the kitchen to the dining area, and as we all fought to get through the one door, a kitchen knife with a huge twelve-inch blade came flying through the serving hatch, across the dining area and embedded itself deep in the folding panel of wood, two inches from the culprit's head.

There is one thing you learn quickly on a station. Don't mess with the cook.

One morning, when the cook came in for work, we gave her a telephone number and said she had to contact Councillor C. Lion urgently, as he had been on the phone to her an hour earlier, saying she had to get in touch right away.

Looking worried and apprehensive, she tried to get out of us what it was all about, but we told her we didn't know. So she got on the phone there and then and called the number we had given her. When she got through, she asked for Councillor C. Lion, saying who she was and explaining that the councillor had contacted her an hour earlier.

All of a sudden, she heard a huge roar of laughter and the person on the other end of the receiver asked if she knew where she was calling. It was at this point that we all made our apologies, saying we would leave her in privacy, and then scampered. We had only given her the telephone number for Edinburgh Zoo. She was furious and vowed to get us all back for what we had done, and she did.

It was in the days before the issue of sleeping bags and we had blankets and sheets to cover our beds in the stand down periods of the night shift. We used to make up the bed in the bay and leave it there for the following night shift, nice and tidy. Unbeknown to us, the cook had got to the beds in the daytime shift as it was not uncommon for the light to the bay to be left off so that others could sleep.

Having reached my bed I proceeded to pull back the top sheet and blanket and jump in. As I stretched my legs out and threw the blankets back over me, there was this almighty rip as my feet went straight through the sheets. The cook had made me a French Bed where the bottom sheet is pulled half way up again and folded over the top of the blankets to make the bed look normal. Mind you, I fared better that night than some of the others who just couldn't throw the sheet back at all. The cook had sewn the top sheet to the bottom one making this impossible.

Not content with that, the very next day without telling any of us on the crew, she had enlisted the aid of the new station cleaner who had only started the week before. In addition, the secretary of the station, the one who we had tied to the flag pole months earlier, was by now well settled down in marriage, and only too willing to get us back for that episode. Between them they sat for hours with paper punches and coloured paper, making holes in the paper and producing loads of coloured confetti in return. Since we were preparing to start the night shift that evening, we saw none of this.

We had no sooner taken over from the day shift, who by now had departed the station, when we received our first call of the evening to a reported fire in a public house just down the road, which meant we had to be quick in getting our gear on. We grabbed our fire kit from the pegs and drove out of the station trying to get our gear on in the back of the appliance en route.

As the first one scrambled to get into his boots and leggings, which were wrapped round the boots for speed, he got the shock of their life as confetti flew everywhere in the cab virtually obliterating everything in sight. Everyone seemed to be in the same boat until I looked and found my boots had no confetti in them. I was feeling pleased as punch that she had left me out and slipped into my boots and leggings.

Meanwhile the rest of the crew, not being able to get into their boots had opened the side window and were gaily pouring confetti out of the window all the way down the road. That was until one of them reached over to the window to pour it out and dropped his boot from the open window. Screaming at the driver to stop, he had to run back a hundred yards to pick up his boot.

I, on the other hand, was well advanced in dressing as far as the others were concerned. Having got my boots on, I proceeded to put on my tunic and helmet and reached for my neck scarf, which was in my tunic pocket. That was when I discovered that I couldn't get to it, as for some unknown reason the pocket containing the scarf just wouldn't open, no matter how hard I struggled with it. And neither would anyone else's.

As we were to discover on returning to the station, the three ladies concerned, the cook, the cleaner and the secretary, had got together as one and hand filled every pocket bar one on each uniform with confetti, sewing up the pockets afterwards.

When we did manage to unpick the sewing and empty out the confetti into the bin, we discovered that the last pocket on each, which was also sewn, contained a hand written message saying "Gotcha." Feeling extremely pleased that at least my boots hadn't been filled as well, I proceeded to unpick my last pocket only to find a note there which stated, "And you too, Whistler. You're not exempt."

The moral of the story is, of course, never unleash the wrath of a woman, for it will always backfire on you men, no matter how long it takes.

Chapter Eight
THE HIGH AND MIGHTY
CONSIDERATIONS OF INDUSTRY

For a fire fighter, there is only one reason for him or her being there - to put the fire out and make sure it gets no further than it already has. So it usually goes without saying that the proprietor of any premises affected is only too willing to oblige in doing anything asked of him in order to get that fire put out.

That is until it comes to dealing the very big boys where it will cost them real money to close a particular plant down, or interrupt a continuous production flow. In these situations there is always an extra special highly qualified expert to advise differently, and the following two anecdotes relate to such incidents.

No matter what brigade you serve, there never seem to be enough men and women to go around at times of sickness and holidays and if you have more than the minimum crew to man an appliance, the danger is that you might well expect to be assigned to the crew of another station which is short of manpower. This is called an 'out duty' in the trade.

On one such Friday evening we were one up on the crew of our station, but at another station, they were one short. As it was my turn to go to out duty, I was dispatched to a major town on my brigade patch to cover a station not normally my own.

After only a couple of hours on watch at this station, we received a fire call to a large petrol-chemical plant in the next town. People were reported to be still inside the plant and had not been seen since an explosion had occurred on the production line.

I was on the first of two motors from that particular station to arrive and a third was on its way from a village on the other side of the plant. As we approached, we could see a fireball leap about one hundred feet into the air, and it was reported to us at the gates of the plant that four men were missing after the explosion.

We drove into the heart of the plant and immediately started our main priority which was locating these men. We took very little time in locating two of them, as they were what one would call the walking wounded. So we immediately dispatched them back to a waiting ambulance for treatment after questioning them briefly about the location of their partners. Both replied that they had not seen them since the explosion, but one had been working a crane and the other assisting.

On trying to move forward and locate the other two lost men, we were constantly driven back by the amount of heat given off by this very large processing tank which had exploded, bending and twisting metal all around and by the huge fire ball which now accompanied the explosion. The fire was constantly being fed with fuel which was burning through the pipe work alongside the tank.

With the help of other crews who had arrived by this time, it was decided to cover the tank and accompanying fireball which at times was now shooting to heights of one hundred and fifty feet. To achieve this we decided to use ground monitors. These are unmanned jets placed on the ground to surround a fire to cool it. We could then get on with the job of trying to locate the two missing men.

We eventually found them some twenty minutes later. Unfortunately, they had caught the full blast of force of the explosion and never stood a chance of surviving.

Worse was to come. We requested the chief engineer to shut down the plant immediately. He refused to do so, saying that the plant, although cooled down by the water on it, was still too hot to shut it down because of the danger of further explosions. When, a few hours later after further cooling, he was again requested to shut it down, he took the advice of his plant manager regarding the cost of shutting the plant down, ignored our advice and refused.

The fire burnt until the following Sunday afternoon, when eventually the Plant Manager after a lot of consultation by telephone to people higher up the ladder, shut the production line down and the fire was eventually extinguished as it no longer had fuel to burn. This was not to last long though.

The following Sunday morning, we were on night shift from the previous night and were nearing the end of our watch. Some of us were up, some preparing to rise as the night had been quiet, when all of a sudden on my own station some twelve miles away from this plant, there was this tremendous shock wave which shook the station and rattled every window in it.

We turned and looked out of the station window in time to see this huge fireball rising some five hundred feet into the air. It could be seen over twelve miles away. We knew then that we were on our way and hurried downstairs from the dormitory to the appliance before the bells sounded to turn us out.

As we journeyed the twelve miles to the plant and were still some four miles away, we could plainly see a fireball rising some five hundred feet into the air. It was an awesome sight.

On arriving at the plant gates, we were amazed to find that we were the first motor to attend. We later found out that the others nearer were attending a major house fire, but we did not know this at the time. On turning into the plant complex, the first sight which confronted us was a plant fire engine flying out between the rows of tanks with blazing away. The hose had caught fire from the heat that the appliance had to endure in the moments when it had gone in to try and shut off the supply valves to the affected area.

Our first job was to make this works fire appliance stop so that we could put out the fire trailing behind it. This only took a few minutes after we uncoupled the hose from their appliance. As we were putting out the fire on the hose, two other works fire fighters came running out of the blaze in silver heat suits.

They had been sent in to shut off the supply valves, and the works appliance was meant to keep them cooled whilst they worked. With the sudden departure of the works appliance because of the trailing fire, they were left with no cover, so they legged it. Suffice it to say that they were very hot and their suits were on fire as they came racing out of the affected area. So we had to put the men out there and then and get them out of their very hot suits before we could go any further.

With this accomplished, we were making our way across the complex to what seemed to be the greatest affected area of the fire, when we heard further appliances arriving. Were we all grateful for that. We turned the corner to where we believed was the middle of the fire area yet far enough away to be safe for the appliance. We could not believe what we saw.

There, right in the middle of the plant we were confronted by a fire at least a mile square. I think that very few of the tanks in the plant were unaffected and even those had blown their vents and had huge fireballs gushing from them. It was as if we were in a war zone where dear old Arnie had pressed the plunger, just like he does in his movies.

It was dreadful not knowing where to start, with only one motor and five men at that point and the water we were carrying was of no use in fighting a running fuel fire.

We needed foam, and lots of it, to smother this fire, and we weren't carrying enough. So the best we could do was order up what barrels of foam they had on the plant, plug them into one of the many fire hydrants they have around and make our way steadily forward carpeting the fire with foam to smother it as much as we could. Other crews were doing much the same in other areas of the plant and it wasn't long before we were dangerously low on foam stocks.

Back at control, they were arranging for further stocks to be supplied from all other brigades as far away as Lincolnshire. They were coming by the lorry load with police escorts. Such was the size and ferocity of this fire. Yet again, we were advised by those on site that two men had been caught in the middle of this blast and we were trying to make our way there first, to the place where they had last been seen before the explosion.

The more we were able to move forward by carpeting the fire with foam, the more devastation came into view. Twisted metal fragments and frames of still burning tanks littered everywhere. Works vans that had been inside the area of the initial blast were twisted and almost unrecognisable. A specially reinforced control building had completely caved in on its front and was just hanging there.

This type of fire is called a 'running fire', where the liquid on top doesn't actually burn on the top but the fire is a few inches above the liquid itself. This is because the fuel is lighter than air which makes the fire burn with a blue flame like that of methylated spirits. It is able to travel very quickly because the fire is freely above the liquid, and it runs in the form of a "V" shape faster than you can keep up.

On moving forward with this very large foam branch, I was on the front and through the fire we could be see some five forty-five gallon drums. We were advised by the works fire officer that we had to put foam on them, there and then, as there was a special kind of fuel stored in these barrels which was very volatile when exposed to fire. Having this rather large branch like a hero I dragged it further forward, along with my colleagues, so that we could throw the foam directly onto the barrels which were stacked three on the bottom and two on top.

We had only put foam onto them for a few seconds when all of a sudden, like the space shuttle taking off, there was a flash of flame and white smoke.

I think I saw the last of them take off momentarily as it became a mere speck in the sky hundreds of feet up in the air. Where the barrels had been, they were no more, just a blank empty space. We never did see them again, but somebody somewhere got a fright that day as they came back down to earth.

Every time we knocked this fire out, it would re-ignite elsewhere on the fire ground, coming straight back to where we were and at such a speed that we had to run for our lives to avoid the blue "V" at the leading edge. After a few hours of tackling the fire and it re-igniting so often, we organised safety officers to walk round the perimeter of the fire and warn the others inside the area of any impending re-ignitions that would endanger the men inside.

Our signal to them was to be continuous blasts on our whistles to get them out immediately. To this end, my partner and I were walking round the perimeter of the fire and on reaching the far corner we turned round to see what the cause of the commotion was behind us. As we turned around, we could see two of the plant workers leaping like kangaroos out of eighteen inches of water and going in two different directions. Both were clad in the same type of overalls. The "V" of the fire shot past one at thirty miles an hour, but the second was not so lucky. It caught up with him and went right up his posterior, and I had to grab a hose reel from the side of the road and cool him off quickly.

Realising the danger to the rest of the fire crews inside, I immediately blew on my whistle continuously as loud as I could to get them out. My partner should have been doing likewise, but when I turned around again he had completely disappeared and was nowhere to be seen. On taking a head count when we got them all out, the only one missing was my partner who still hadn't shown up.

We had three turntable ladders with us that day, using two as towers to pour water onto the fire from above. The third was in reserve up by the works canteen about three quarters of a mile away from the fire. It was here, behind this third turntable ladder, that we eventually found my partner. He explained, that when he saw the fire coming towards us, he concluded that the whole plant might blow because of it, in which case he wasn't going to be there when it happened. So he had legged it to safety there and then, and at the end of the day you can't really blame him. A dead hero is no good to anyone as I explained earlier.

This type of fire is known as a major civil emergency, and the plan of action is worked out in advance. In these situations, the fire master would go to a safe area and his deputy would run the show from the fire ground, which was what happened that day. Such was the force of the initial explosion that created the fire that the base plate of one of the storage tanks, which was four feet wide and six inches thick, was thrown hundreds of feet into the air and eventually ended up embedded in the wall of another canteen on the plant. It had sliced through a brick wall and was sticking out of the wall, half in and half out, miles away from the explosion.

We should have been relieved at ten that morning by the oncoming crew but by four in the afternoon, we had still had nothing to eat. There were no relief crews and it was a bit disconcerting as we looked over to the main road running through the middle of the plant and saw beyond the perimeter fence all these TV crews setting up. Worse still, we could see twelve 'ECON' trucks that had never been used parked in a line at the side of the road. ECON trucks are snow ploughs with the big blade on the front, and the back filled with sand.

It took us four days and nights to put the fire out, but such was the initial bang of the explosion that it took out every window in the town in a two miles radius, and everyone for the first two miles had been evacuated because of the danger of further explosions.

It took another two years for the truth to come out. It emerged that part of the major emergency plan was that when the particular fire master had used up every available crew from one watch, plus the part time men, as well as several other vehicles from other brigades, and it had still not worked, then he was prepared to shove the plant and us into a nearby river to save the town. He would have levelled the plant to the ground.

At least according to this plan he would have had a full three other watches left to carry on with, at our expense. We did find the two unfortunate men some hours later. They were located under a wooden framed wall and in water. They had been hurtled some two hundred and fifty yards from the explosion centre and never stood a chance. How sad these things can be.

A senior officer of one brigade once said to me, "I would love to help you son, but it's bums on seats." How right he was that day, and about the job itself.

It doesn't always pay to bear the cost factor in mind when fire is involved on such a large scale.

This was something the company concerned found out on that first day of operations, when you take into account the distance fire fighters had to come to fight the fire over the total area concerned, and the speed at which a re-ignition occurred time after time. Then you add hoses, brand new specialist equipment, vehicles and everything that disappeared in one huge flashover. There was not much left other than charcoaled remains after we had smothered the fire with foam again and again. When the cost was eventually added up, many millions of pounds had been lost in a few moments.

Thinking back, so many men were involved in fighting this fire at one point, that when you looked at the total number of fire appliances all lined up there one behind the other, a safe distance back from the fire, it was as if someone had taken every fire appliance in the U.K. and parked them all nose to tail in a long, long line.

Chapter Nine
THE ROAD TO YONDER LIES

Fire fighters rarely enjoy the luxury of a happy steady relationship, and do tend to get through wives in their careers. One can imagine why. Fire fighters are brave but you can understand the pressure for wives (and husbands) as they see their men and women off to another shift, at the back of their minds wondering if this might be the last. Thankfully, such is a rarity, thank God, because of training and professionalism on the job. Nevertheless the nagging doubt is always there with the wives and husbands of fire fighters, and a lot of unspoken pressure is applied which very often ends up in the break up of a marriage.

After over a decade and just about every conceivable type of emergency covered on the same watch, I embarked on a new marriage and after a transfer request and interview I was off to a different country, a new challenge, and a different way of life for my new wife and kids.

We drove down to our new station from Scotland overnight, as the distance was quite a considerable one, and with the morning sun rising over the motorway, you suddenly realize when you leave Scotland, how very cold and damp the climate can get there. Still without a doubt, it is one of the most beautiful countries in the world for scenery, not to mention the nature of the people who live there.

Having been one of the lucky ones to see the world several times over in my military career, here I was back to where it had all started near enough twenty odd years ago. The only difference was that this time in my car I was taking with me a new wife. My first marriage had ended many years ago, with me losing touch with my children at their most impressionable age. My wife had disappeared in the middle of the night with my (supposedly) best friend. These things happen in life, but when they happen in that manner, they still leave you feeling as though someone has twisted a knife in the pit of your stomach. This is especially true when you feel you are doing your best for everyone, including so called friends, and I was to wait eleven long agonising years before I was to rediscover my children's whereabouts.

I had gone for some years believing I was an independent so and so. You begin to believe this after a few years of being on your own. But here I was married again with three of our four children in the car and once more looking to the future.

And as the sun came up over the motorway, it was the start of one of those beautiful days in early spring. It wasn't long before the temperature was up to twenty-five degrees. Twenty-five degrees! In my whole career I'd never seen twenty-five degrees in Scotland, sometimes close as it might have been. But it was a glorious day.

We arrived at our new station, myself, my wife and three of our four children. The fourth child had stayed behind in Scotland because she wished to live with her Grandma and at sixteen she had reached the legal age of consent in Scotland. I had left my old brigade early so we could all move in with the minimum of fuss using up my few remaining days of leave. Our belongings were somewhere on the motorway behind us in the removals motor and would arrive later on in the afternoon. We picked up the keys to the new home we were moving into, and opened the front door. But we hadn't bargained for the state our predecessors had left the place in. By then, of course, they themselves had moved on to fields anew.

One of the reasons I had moved from my old station was because, although it's my country, Scotland has many areas of black spots with a lack of posts. Unless you are highly educated, or have in you that certain area of skill you require, it is very difficult indeed for young people to get and hold down a job, assuming they are lucky enough to find one in the first place. I sympathize with all young people there today.

Having seen all four of my kids unemployed for something not much short of eighteen months, I was intent on them being given the chance to set themselves on the right road in life, and through hard work to give my new wife the greatest opportunity that I could provide her with. However I wasn't kidding myself that it would be any kind of picnic to integrate in my new post when we first arrived.

I need not have worried, however, for within two days of arriving in our new chosen town, all of our children found employment, and to this day this is where they all remain. They have not looked back, and therefore we feel that, as a family, we have nothing to regret in making the decision to move on.

At home back in Scotland, it was not unusual for all the family, including all their children who were too young for school then, to turn up first thing in the morning, and leave only when the tea meal had been eaten. Then they would all go home and return first thing in the morning to start all over again. This meant that my wife would spend all day in the kitchen looking after them all including the mother-in-law, now deceased, with whom I had a wonderful interactive relation. We got on amicably, but mother-in-law would order the kids around. "You do the tea. You make the sandwiches. You fetch my slippers." When it came to my turn, I would sharply remind her that Adolph Hitler died in 1945 and I would like to keep him buried and I'd tell her to get off her bum and do it herself.

Such was our relationship, mother-in-law and myself, that she would have got up and clocked anyone in the family round the ear if they dared to say that to her. But not me. Mind you, she never meant harm by this at any point. She just wiled her way through life by charming everyone into doing the things she wanted doing without doing them herself. And our kids without exception loved their grandma. She would give her last fifty pence to them one week, and borrow a pound back the next, such was her nature. So God rest her soul. She never did get the opportunity to have much in life, and by the time my wife and I did get on our feet, she'd passed on before the real benefits could come through. We did try to manage to give her those things, as it was God's will. But I feel privileged to at least say that in her final days, my wife and I were still looking after her, and in her final hours in hospital, we were still there.

When I get back to thinking of her endearingly, I have to say that my mother-in-law was quite some character. She once came over in the car with my sister-in-law to stop in our town for a holiday and, I have to say, my mum-in-law loved her Bingo. In one week, she was out at Bingo so often, every afternoon and every night, that when I tallied up what she had spent in the place over the whole week, it came to an amazing two hundred and ten pounds. The money was provided by myself, for I would have given this lady the shirt off my back if she had asked.

When my sister-in-law came to collect her and take her home, she turned round as she was getting into the car and said to my wife and myself, "Thanks very much for the holiday. I really enjoyed that. I think I'll come back for another week." To which I replied, "No, you stay where you are and I will visit you. It's cheaper, Mum."

Believing that we had pulled a flanker on the family by moving away to the other end of the country, far enough so as not to have to go through this daily routine, my wife excitedly decided to call her sister to tell her we had got the job, only to find that her man answered the phone instead. He eventually stumbled out that his wife had left him a fortnight before as a result of his indiscretions, and was now living with the other sister and mother-in-law, plus all the children, in a caravan park only sixty miles from us in our new home.

Somehow inadvertently, they had all reached the other end of the country before us. So don't ever think that you can be lucky enough to run away from that kind of situation, because somehow you will still end up the same boat, only further down the road.

However I have been lucky enough to have a close relationship with all my in-laws and out-laws combined, and a good one too, and at least now we had sixty miles of distance between us. So they had to make the effort now, if they were all going to come over every day, and of course that did not happen. Mind you, they would as often as not come to visit whenever they were able to get on their feet and arrange their own little car to get them back and forth. And likewise, we did visit them often in their caravan park.

On one occasion, in a very cold February, we visited them, only to find that it was so cold that the water pipes had frozen and we found them all huddled up in front of a one bar electric fire. We had to take them all home with us before they got pneumonia. We had no choice. So there we were looking after the whole lot again only for a few days, yet this time for twenty four hours a day instead of them going home after the meals were eaten. So I think the moral of the story here is so long as you have family around you, you can never run away.

Anyway, getting back to the story, we had just got the keys to our new home, although it didn't look new, and it mostly took my wife to get it up to the standard we wanted as like myself, the children were working full time. I would leave my wife at home to sort the place out during the day, and we would help on returning home in the evenings. But it took a full six months to sort it out, such was the condition it had been left in by the previous occupiers, who had patched up the place to a standard they had felt comfortable with.

My wife was in tears with the place after that six months were over, and had virtually become a recluse. Not speaking to anyone for all that period of time had made her withdrawn, and she needed a holiday away from it all to get back to normal. First of all we went to friends we had previously met on honeymoon in Italy and who had invited us to their home in Wales for a few days. From there we went to a caravan park in

Wales on our own, and gradually I saw the first signs of the wife I knew begin to return.

The one big change I did notice was the difference in nature between the Scottish people and the English. In Scotland, families tend to have only one working adult, and they are more laid back, more friendly in nature than the English.

Before departing my old brigade, however, I can recall a certain TV company who came to call at my old station.

A pompous producer with a double-barrel name was making a programme on water, how we get it, how it is cleaned up for use, and what can be done with it when we have it. With this in mind, he decided to make part of this programme on how the Fire Brigades use water.

After some discussion with the chief, it was decided that those of us on station would play hosts to the film crew for the day and that they should film us going through our normal daily routines. But because the programme was mainly about water, they had ideas of what they wanted and asked us to go along with them. So we agreed to do that for them.

As part of this, they wanted to film was how we got the water from the ground to the fire appliance. So we agreed to get the youngest member of our crew and let them film him lifting the hydrant lid in the drill yard, then screwing on the instant coupling connection stand pipe. He would then insert the hose into this and run the hose away fitting the branch ready to receive the water from the ground hydrant. Then one of us would turn on the hydrant to show what happens next.

The camera man got into position and on countdown said "Go" and our youngest got to it. However, what none of us had bargained for was that David was a whiz kid at doing all of these things. As he reached the end of the drill, David, now panting like an old man, discovered to his horror that he had gone at it so quickly that the camera man had only kept up with him on the first bit. So he had to do it all again. But again he was too quick so he had to do it yet again, and again, and again, until he was so knackered he could barely stand upright, far less do another drill. The rest of us instead of helping, were all curled up in fits of laughter at the inability of the camera man to keep up.

Another idea they had, was to show the appliance turning out of the station as if it were going to a real call. The plan was to get the camera in low and close to the main front door and Tommy was to drive the appliance out past the camera man as close as possible for dramatic effect.

That day we had been having problems with the pump leaking and had to connect a short piece of hose to the pump outlets at the back and out of the roller shutter doors because of the water that was accumulating and making the tiles in the bay slippery. So the coupling of the hose was connected and the other end stuck out of the roller shutter doors. Then the door was lowered so that it was only open a little above the level of the other end of the hose.

However, because the TV programme was introduced by a very charming female presenter, their idea was to put her on the fire appliance. First they wanted to film her on getting into the fire appliance, disappearing with it, then saying a few more words when the fire appliance returned to the station, with the camera man taking the footage by squatting down at the bottom corner of the door as the appliance went past.

Someone had to get off the appliance to let her get on as there wasn't room in the back of the cab. So I duly stepped off and let the girl get dressed up in my fire kit. Then she stood on the steps of the appliance where she said her pretty little prepared speech on film as planned then got in, closing the door of the crew cab behind her.

The camera man got into position on the bottom right hand corner of the door to film the appliance as it left the station. Then he joined it and resumed filming from inside the cab as we were going down the main road outside the station driving around the round-a-bout at the bottom and returning to the station. This would give him some good footage.

From his position at the door, he could film the blue lights being turned on, and the horns going, but to do this from that position he had to crouch right down and balance himself on his feet. The motor started and Tommy drove off at speed past the camera and out the door.

At this point I was standing to one side of the appliance and to the rear of it where I would be out of the way of the camera. As Tommy drove out of the station, there was an almighty crash and the door at the rear of the station suddenly came flying off its roller and guide at the bottom. The hose coupling came whizzing past my nose like a bullet only a couple of inches away. In their haste to film, everyone including myself had completely forgotten about the hose coupling we had shoved out of the rear door earlier At the same time there was an almighty scream as Tommy ran over the cameraman's foot on the way out.

As I looked out towards the front door, my face chalk white by the near miss I had just had, the poor unfortunate cameraman was jumping up and down at the door holding his foot. All the blood drained from his face and he too was white as a sheet at the near miss.

However after a minute or so he regained his composure and hobbled onto the fire appliance to take the rest of his footage from inside the cab as Tommy sped off down the road with them all on board. The appliance was away for two or three minutes allowing me time to fix the door back into its runner and get it working so that we could raise the door before they came back.

When they returned, Tommy drove the appliance round the back of the station and into the bay to the position where it first started out. Now was the time for the rest of the plan to be put into action, with a short speech from the presenter at the end standing on the steps of the appliance. So I went to open the door to let the cameraman out of the cab so he could get in the correct position for the presenter's speech.

What met my eyes as I opened that cab door will always be the enduring memory of my old brigade for as long as I live. Inside was this female presenter facing me, with her fire helmet tilted at an angle, sitting bolt rigid upright in the seat, with her arms straight by her sides and her face chalk white. No matter how hard she tried she was just unable to move. Next to her was our cameraman who had been thrown all over the cab by Tommy's driving, completely unable to get a single inch of film inside. He also was chalk white, and the two burly fire fighters next to them were rolling about on the seats with laughter, tears running down their faces so much that they couldn't stand up, while in the front was our sub officer physically punching Tommy for his crazy driving.

Instead of the planned speech on the steps, she turned to the producer and said "Never again. I don't care how much the job pays. I will never again get on a fire engine." Seemingly Tommy had gone off at such a great knot of speed to give a realistic impression that they went around the round-a-bout at the bottom of the road virtually on two wheels, with tyres screeching, and then shot back up the road at high speed to the station.

That was enough. While the driver and sub officer were still beating up Tommy, I along with the rest of the crew were rolling about in fits of laughter on the mess floor. The TV company sent us a letter telling us that the programme would be broadcast a few weeks later. We all filed into the recreational room in good time that day and switched on the TV to watch.

It had taken all day long to film us, and when the programme did come on and our section was shown our share of the whole programme lasted less than thirty seconds. We reckon that the film was so unusable that most of it was spliced out. What an endearing way to leave a brigade.

Every fire station has its characters and its individuals and David, our youngest joiner to the watch, was certainly a character.

He was the only joiner on our watch who came from the city, and I remember him one day in the back of the appliance asking the sub officer, "What are those things there with big brown sticks, which are thick at the bottom and have those green things all over the top of them?" The sub, being perplexed at the question, asked what he was talking about. He replied, "Those things over there by the side of the road."

The sub, now realising that they were trees he was talking about said, "They're trees!" To which our David replied, "Oh! Is that what they are? They don't have any of those in the part of the city I come from, only concrete."

All of us had homes to go to at the week-ends but when we asked David what he did Friday nights, he answered that he lived in a telephone booth standing upright, because after spending the night drinking in a local bar this was the only place he didn't fall out of as he was usually so drunk.

The first few months of joining any new brigade are taken up by being gradually introduced to different equipment. There are rules that prevent you from driving a brigade vehicle before you have undergone a course and all emergency fire appliance drivers are also required to undergo refresher driving courses every three years, and rightly so. You also have the task of learning the ground to which you have attached yourself. So your first few months are not so much spent on station as on course after course that you must attend in order to acquaint yourself with your new brigade. But it still comes as a bit of a shock when you learn you cannot touch the breathing apparatus compressor until you have done the course, even though it is something that you took as automatic as you had been doing it anyway for the last ten years in your old brigade.

However, having undergone all of these introductory courses, it was now time to join the new station properly.

Chapter Ten
BREAKING NEW GROUND

As I have previously mentioned, when you join a new brigade, many of your first few months are taken up with getting to know those you are working with as a team, learning their methods, attitudes and requirements. To this end and in order to blend in and be part of the team, it is essential to carry out a lot of training with them.

It is also necessary to spend a lot of the time out on your station ground, getting to know the roads and streets that you will be required to cover on your patch. Having gone from one kind of shift work pattern to a completely different type, I, like the others on the station, would have to cover a lot more than I had previously done. However, it wasn't long before my new colleagues' particular brand of humour began to emerge, and I soon fully adjusted to it.

Like everywhere else, we often didn't have enough men or women on the station every day.

As a result of the undermining of posts on the station, as others transfer away and their places are not immediately filled, for the first eighteen months I found myself as only one of two drivers on the station. Then the other driver, through sickness, took temporary long term promotion to leading fire fighter. So I found that, as he could not drive and be in charge of the crew at the same time, it worked out that I was the only driver. There was the odd day where he would volunteer to take the job of driving, which would allow me the opportunity to sit in the crew cab and fight fires, which is what I had been more regularly accustomed in my old brigade. However, in any brigade, you just do what you can to ensure that the appliance gets out of the station when it is needed. You do what is necessary, not what you would prefer.

Junior doctors fight to reduce the seventy-two hours a week on duty, but we fire fighters were now putting in one hundred twenty hours on station or on standby at home ready to dash in to the station at a moment's notice. This makes the doctors' hours look ludicrous. We had five long days and nights before we could get off duty, with three or four days off afterwards to recover before starting all over again.

Such was the shortage of manpower on station that often the answer was to get one or even two men from other stations later on in the morning. Frequently, at least one of these men would not have long been in the brigade, and it was these poor souls who very often bore the brunt of our humour.

Prior to my arrival in my new brigade, the lads on station had found the frame and wheels of an old post office worker's delivery bike round the back of the station.

The bike was very old and rusted through. It was the type with the old iron brakes you pull up to the handlebars. With a bit of minor surgery, they had got the tyres and accessories working. One day, they were messing about with it in the drill yard in front of the station, when a senior officer drove into the yard unexpectedly and caught them at it.

When they should have been working, there they were all out front taking turns in riding this rusted heap, with tyres only half inflated, and peddles lying at angles very different to how they should have been.

When the senior officer saw this, he immediately asked the sub officer in charge that day to explain what the crew were doing with this bicycle. The sub officer, knowing that his neck was about to get stretched if he didn't come up with something quickly, replied that this was the pedal bike from the brigade's hose laying motor, which had broken down and that the crew were trying to fix it.

The senior officer, taken in completely by this story, looked at the bike, and was horrified at the state it was in. He immediately went into the station and telephoned the brigade's stores group to send a general duties driver over without delay to collect the bike from the station and have stores do something with it to improve its appearance. He ordered the crew to label the bike for stores to pick up and as the senior officer departed the station, the crew all went back to their normal duties. A few weeks later, a station not too far away had this same bike delivered to its hose layer.

The bike had all been sanded down and fitted with new brakes. Cables had been replaced, pedals straightened, and the frame had been painted red and white in fire brigade colours. But the best part of all was the finishing touch. A brigade badge had been placed on the frame of the bike beneath the handlebars proudly depicting the name of the brigade.

Luckily, our sub had been in touch by telephone with the sub officer of the other station explaining the circumstances and asking him not to let on that anything was out of place when he received the bike. When it did arrive on station, the other sub officer simply said to his crew that the hose layer bike had arrived back from stores and would they clear the back of the hose layer motor sufficiently to get it on. It remained there for many years.

Many recruits on our station, new from training school, didn't have a clue what they were doing. Theirs was the first station they had been attached to since they arrived in the brigade. So we as a crew had to do something about it when we saw one of these new recruits arrive on our station for the day.

Because of the short distance to the station, a lot of the lads would cycle in to work. So every day there was always the odd bike lying around. The sub officer would ask the young lad, just to prime him, if he had been on the hose laying bike on his station, and they would always reply, "No, but I have seen it there."

This was the very answer we were waiting for, and our sub would explain to the new lad that, before he or she was allowed to ride it, they would have to undergo a day's training course.

We had tied together two metal pieces of pole with string, one long, one short, and tied an old pair of shopping trolley wheels to the short piece to represent the axle of a so-called 'hose laying trailer' which the station was supposed to have. Then we told them that the length was the same as the hose laying trailer. It was attached to the seat of the bike for pulling by a piece of string.

The sub officer would then say, "Don't worry, son, I'll check that it's alright with your sub whilst you're here," and then ask the young lad to go out in the yard and help the crew set up the course. Naturally, we were all more than enthusiastic about that. The crew would play the part amicably, laying out road cones at certain distances for this young lad or lass to ride between on the bike as part of the course they were to undergo. And we always made sure that the person was kept busy helping.

The sub officer, meanwhile, was supposed to be on the phone to this young persons sub. In reality he was waiting in the station until he could see the course was all set up. He then came out and ordered the crew to attach the ram shackled frame of a trailer we had made to the seat of the bike, and ordered us back inside supposedly to get on with the work routines.

We would all hang about inside the station, out of sight but watching, as the sub would dutifully get this young one onto the bike and explain to him or her that the purpose of the course was to get them riding safely with all the right movements required to pull a trailer. To this end he would suggest that the person got on the bike and rode in and out of the cones we had set up, just to get the feel of the bike before the actual exam began. And, of course, they were always keen as mustard.

When they had done this two or three times, the sub would then ask if the person felt comfortable now and was ready to begin the required test. As before, he would make them go in and out of the road cones and when they had done this a couple of times he would introduce more cones gradually, so that the person concerned would not suspect, thinking that it was still part of the original course.

Then, as part of the test, he would ask them to show their prowess by demonstrating their knowledge of the highway code. They would do this by raising their arm as a signal as they drove between each cone. Having got them this far, to make them think, he would intervene and ask the person what he or she would do if someone was in the way. The usual reply was "I would ring the bell, sub." He would then ask them to do this the next time as they went through the cones as well as raising their arms, and he would send them back to the beginning of the course to start again.

You can imagine our reaction as the poor victim went in and out of each cone on his bike, arm up in the direction he was going, changing arm and ringing the metal bell on the handlebars as he went through. In the station, four other blokes were screaming and rolling on the floor in fits of laughter watching this all going on.

It was a sight to see and after another ten minutes, during which the rest of the crew could not stand up straight, the sub would call an early halt to the test, saying he had seen enough and that the person had passed satisfactorily. He would call the crew to come out and clear up whilst he spoke to the candidate in the station.

On arriving back at his own station the next day, our sub had primed the candidate to mention the bike test to their own sub and ask if they could have passing the bike test added to the list of their qualifications, which they all did. It came as no surprise that, when on finding out that they had all been duped so well, we received a telephone call from the individual telling us to the effect that we "all had no mothers".

The railway line above the fire station was very close to the railway station itself, and every train on arriving at the station or departing would blow its horn. We decided to invent a training course for new recruits to test them on their prowess as railway warning experts. Every fire appliance carries horns to respond to a train's horn on a railway track when we have to work there. They are carried in the lockers and used to warn both the crew and the train driver that his approach has been spotted, and the crews have been warned. They sound exactly like a train's horn when blown and the procedure is to respond by blowing twice when a train driver blows once from the train.

Having got these, we were determined to make full use of them, for it is not every day you go on a railway line. We would ask the recruits if they had been on a railway line on their station ground, knowing full well that they had not. Then we would get them round to the locker of the appliance and ask them to show us how to blow the horn in case of oncoming danger from a train on the track. Full of keenness, as always, they would show us how to blow the horn after a couple of shots at it, as it took a knack of knowing how to get it to work.

We had a drill tower, which at its top stood more or less level with the railway line above, and this was where we would introduce the challenge. "Yes, but can you do it to the correct timing, so that we can get crews clear of the track before the oncoming train strikes them?" We would wait a while and then tell this young person "It must be your lucky day. We have just received a phone call from Rail Track informing us that today they are carrying out a test with all train drivers leaving the nearby station. The drivers will blow their horns on leaving, and they want us to have someone respond by blowing twice on the railway warning horns. You will need to be on the roof of the drill tower for the train driver to check if he can hear them."

Duped, keen as mustard, and believing they were helping the test, they would be asked to climb to the top of the drill tower and take the first watch of two hours, when someone else would be sent up to relieve them. Meanwhile we sent another lad round the back of the station with the other railway warning horn. He could plainly see the track from the back and he would wait until a train approached the station. Then he would wait a moment until the train in question began to pull out of the station and <u>he</u> would blow his horn as if he was the train driver. Now the young lad on top of the tower had been instructed to wait for the one blow and reply twice.

Mind you, we had already primed him by saying that he wouldn't see the train before he heard the horn, but it was vitally important to the test nonetheless that he responded twice when he heard the train horn.

The lad behind the fire station would decrease the duration of the gap between his one blow on his horn to such short intervals that it became like, "Toot. Toot-

Toot" all the time. Believing it was a train coming, the lad up top was answering every time he heard a horn. As he had to stand with his back to the fire station to see the railway line, we in the meantime had gone up to the first floor, opened the mess window and taken Polaroid pictures of him doing this to present to him when he eventually was called down after a couple of hours, totally knackered. We also took one shot of the man at the back of the station blowing his horn.

Mind you, it was not unknown to get caught yourself with these horns. Since they were black in colour, they were ideal in the semi-darkness for sticking up car exhaust ends so that when you started the car, all sorts of funny tunes came out loud enough to wake the neighbourhood.

Chapter Eleven
THE WAY THE WATER RUNS

Boredom on any fire station can be a problem. No matter how many work routines there are, doing them daily makes everyone more efficient and it takes less time to complete them properly. So it is that you find yourself with a little free time on your hands.

The weekends are probably the worst. You look out of the station window and you see all these people going up and down the street. Meanwhile, you are trapped inside the station waiting on that next call, which might still be a few hours away. So you just have to invent things to pass the time. This is how it was with the hand radios.

The station layout was such that on turning out, the appliance had to come out onto a main road, so it formed part of the front of the station concerned. Many people were passing the station all the time, and we all know how the people of Britain are renowned for their concern for animals and others in distress.

We all knew that as individuals, if we were in trouble, our partners would come to our aid immediately. But would others do the same as a group? We had to find out.

Now our station was a clean and tidy station with flowers and rose bushes planted in the grass border to make it more appealing to the general public. It was not unknown for us to be outside cleaning up the rubbish blowing in from the street and into the garden. It was to one very bushy, very prickly rose bush in particular that we turned our attention, knowing that people would not be stupid enough to put their hands into the middle of it without wearing thick gloves.

We planted a hand radio in the very centre of this rose bush, right next to the pavement, knowing that people using it would have to pass the bush. Then, with the other hand radio, we would go up to the mess room and, keeping out of sight, we would wait for our victim to pass, usually an old lady or gent initially. As they passed the bush, we would say into our other hand radio, speaking very softly so as not to alarm them, "Please help me. I'm dying. Please give me some water".

Invariably the person hearing this would hesitate, wanting to know where the voice was calling from and start to investigate, wondering all the time if they were hearing things, yet still searching around for this voice. When they couldn't locate it right away, we would try to help out by saying into the radio softly, "No madam, over here. Please help me. I desperately need a drink, or shortly I will die!" This is where the general public at large came in. Seeing this old woman searching around to locate where the voice was coming from, they always wanted to help.

So soon she would be joined by another passer by, then another, and another, until you had quite a crowd, all searching the area for this voice. Meanwhile all of us upstairs would remain absolutely silent whilst below they were searching yet thinking this old woman was absolutely off her rocker and was hearing voices in her head.

We would allow the public to search a little longer, and then throw in those words again. Seeing that they were getting closer to the bush by now, we would repeat, "Please help me. I'm dying. Please give me some water", until someone plainly identified that the voice was definitely coming from the rose bush. Having established that it was, they would all look around wondering how this bush could talk. You could see by the expressions on their faces that in the back of their minds they were thinking "Is someone having us on?" Yet they would still look all over to answer the question as to why they were hearing voices from this bush.

Mostly they never did find the radio, for who would be daring enough to stick their hand into the very heart of a rose bush without gloves?

We would remain silent for five or ten minutes while they all discussed it between themselves, and wait just that extra five minutes to see if they could hear the voice again. Not having heard it, they would give up, wondering what had happened and how they got into this situation for the last half hour in the first place. Then they would depart, still scratching their heads in wonderment.

As soon as they had all departed, we would wait and repeat the same trick with the next victim who came up the road until we had another crowd searching all over the place for the voice.

Another trick we used was to plant the hand radio beneath the metal plate of a roadside wastewater drain outside the station. Usually as the unsuspecting victim walked past over the drain and we would quietly say into the radio, "Please help me, I'm stuck," from the mess room above. We always left the window slightly ajar to hear what was going on outside. Inevitably, the victim would be searching around for this voice they had heard and would automatically attract a crowd who wished to help, when the victim had explained what they had heard.

We would allow them to search a little and then say, "No, madam, down here. I'm stuck beneath the drain cover." After drawing them to the drain, we would then engage them in conversation. They would say something like "Are you alright down there? How are you stuck?" We would reply very softly, "I'm not stuck in the drain, but beneath it. Can you please help me get out? I've just come up from an underground mineshaft. I've made my way up this far and got stuck. I heard someone passing overhead and realised that I could get help if I called you."

Without fail, the woman would respond by saying, "Please don't move. I'm going to get help." They knew that they were outside a fire station, so where better to get help. Knowing they were about to ask for help, we would send one man to the main door of the station to await their arrival. The victim would always say, "I know this may sound strange, but I believe there is a man stuck beneath that drain over there, and he is asking for help to get him out."

The fire fighter who answered would always look over to this twelve- inch drain cover and say, "Away, you're having me on, aren't you?" Indignant as they were, they would still insist that we accompanied them to the drain cover. So the fire fighter would always say, "Excuse me, madam, I'll have to go and get the rest of the crew in case you are right." He then got the crew and we would wander over the road to this drain cover.

Having made our way through the crowd that had gathered by this time, we would stand over the top of this drain cover and pretend to listen intently for the sound of a voice and, of course, because we were all round the drain, nobody could say anything. We would even play the part so well that we would ask the crowd to be deathly quiet whilst we all listened for the voice she was supposed to have heard.

We would ask if she knew the name of this person, and as she didn't, we would take it in turn to call out "Hello" and wait to see if we got a response, which of course we never did. After several minutes, we would ask the lady if we could have a word with her to one side and ask the crowd to please disperse, which they did.

We would explain that, intentionally, or unintentionally, the lady was wasting fire brigade time, and would she please not ever do that again, as we had our work to get on with. Feeling about two inches tall and barely able to keep her head up in the street, she would wander away muttering to herself about definitely hearing this voice. In the meantime we had returned to the station and were watching in fits of laughter as she went off muttering down the road.

Only once when doing this trick, did our sub inadvertently burst into tears of laughter at the door when the lady said she had heard a voice coming from the drain. It was almost curtains for us that day, as she clicked right away and said, "It was you lot that did it, wasn't it? I'll report you to your chief."

More by luck than judgement, she never did and, thank goodness, we got away with it.

We were up to mischief a good deal of the time. One day we were bored and wanted to find something to do, so we tied a furry thing, like the tail of Davy Crocket's hat, with a safety pin to the end of a fishing rod and spinning reel. We took it out of the top window, and down across the garden to beneath a broken concrete slab, then across the gap between the wall that leads to the front door of the station. We then waited for our first victim. We didn't have to wait long for our first victim was ideal for the task.

An old woman came walking past, walking her dog. This little terrier was off its leash and trotting in front on its own. The dog spotted this furry thing sticking out from the end of the wall, and immediately came over to investigate what it was, getting over it and sniffing it like they do. While it was sniffing, the lad on the other of the rod suddenly reeled in the line and this bit of fur shot across the gap in the wall at a hundred miles an hour, giving the little terrier the fright of its life.

The dog shot up in the air, with its legs splayed out as it went, and the hair on its back sticking up on end. When it came down again, it flew away from the old woman's side and bolted up the road. The old woman, thinking it was a rat, turned round and bolted up the road herself to howls of laughter from above. Mind you, we had to be careful who we played that trick on because of the risk of a heart attack.

We were forever playing tricks on each other as well. I'm sure you have seen on television how we would to parade on muster to receive our allotted places on the appliance that day. We would be lined up to attention when suddenly, out of nowhere, would come a great stream of continuous water soaking us from head to toe. Someone had brought in a plastic lemon filled with water.

Not to be outdone, next day we all brought one in and carefully filled it up before the parade. As the sub was giving out the duties, we would be giving each other a good old soaking from these things. He gave us a row for doing it, so as one we turned on him and he was forced to bolt through the bay door to escape a drenching.

It didn't stop at that, of course. Someone had seen a trick with Sellotape and a paper cup, where the cup is filled with water and the tape is attached to the top end of it. Fire fighters are experts at picking locks, as you might expect, so it wasn't long before someone picked the lock on a personal locker upstairs, and placed a full cup of water on the top shelf of the locker, attaching the tape lightly but firmly to the inside of the locker door, then re-locking it. The victim, not having a clue, would open up the locker, pulling the door as normal. With one almighty rip, the tape would tear away from the door, pulling the full cup of water from the shelf all over the person's face and shirt, soaking them.

Of course every body was at it then and many, yeah many, got caught with it and then soaked. It didn't stop there either, and it wasn't long before the tops of every door in the station were booby trapped with these things. You would push a door open, hear this almighty rip, and look up just long enough to see the water cup come down and drown you from head to toe.

This had been going for several weeks on the station, and this particular day was no exception. It had been the odd door at first, but on this day these cups were above every single door with the intention of catching every fire fighter on the station. However, it had been going so long, day after day, that we were all careful not to be caught, and up until lunch nobody had been caught by them. So we all went upstairs to have lunch and sit and relax.

I was on my way downstairs to make a personal telephone call, when in through the front door of the station walked a senior officer, who had at one time been the station commander, on his way to the loo. Before I could get the warning out of my mouth, he pushed the loo door open. There was a moment's silence, then all of a sudden an almighty rip as the tape tore, bringing a cup of water down all over his head and completely soaking him and his shirt. He was furious, and shouted at me to get the sub officer downstairs and he would see him in the office right away.

Before I could say another word, he turned round and stormed towards the office throwing back the door to the hallway. Of course there was another rip and he got it for the second time. The sub was placed on six months warning for that little lot, and it was time to stop.

Taking it further, from time to time we used to swap station grounds with others nearby to learn about their station ground. On one such occasion we had gone further North and met the incoming appliance which was covering our ground for the morning. We primed them that if they got bored there were girlie magazines hidden from our cleaner on top of the toilets. In fact, we had attached the sticky tape to the water cups and the moment they lifted them, they got soaked as well. We also told them that we had also left chocolates in locker number one for them, calling out "So enjoy your morning lads," as we left.

Meanwhile while we ran around their station ground, priming others to let us have a free go on their go-karts to see how fast they went, or persuading others to let us have a go on their pilot training simulators, these lads were on our station getting soaked every time they opened a door or touched anything. When we returned to station, the evidence was clear as the floor was covered in water all over the station since we had booby trapped every single door, locker and other material lying about on station.

The final trick was on a new lad on station. He had only been with us a few days awaiting his first course of indoctrination to the brigade. He was at home at the time, not having started yet on the station.

To get the fire appliance out of the bay on call-out, the front doors of the bay were the 'pull-back-in-the-middle' type. We filled a plastic dustbin with water, tied a rope to the handle and gently hauled it up with a rope and through a beam at roof level. With the aid of a ladder we placed the dustbin full of water precariously on top of a partly open door to the bay and locked every ground level door on the station.

We then called the new lad into the station by telephone, saying his new station officer had arrived on station and would like to speak to him right away. Trying to impress his new station officer, the lad drove into the yard in his best suit, and headed for the only open door he could see, after trying all the others and finding them locked.

We were all up in the mess room, watching this going on, as he made his way to the partly open door. As soon as he pushed the door, the inevitable happened, and all of a sudden this dustbin full of water swung half way down the length of the door on the rope and emptied forty gallons of water over the poor unfortunate man's head, to roars of laughter from above.

I swear I saw his suit trousers physically shrink in front of us that day as a result of the sheer volume of water that landed on his head. He muttered something about paying for the damage to his suit, but he relented after a couple of months on station after seeing the funny side of it.

In the eighties and nineties all of Britain suffered an awful lot from shortage of water before the rains eventually came. We were forced to turn off the water supply from time to time, and the local newspaper had announced in headline banners that if our area didn't get rain soon to fill the reservoirs, then the water supply to some areas would have to be turned off for long durations in order to conserve water.

At one point we were doing our normal work routines in a village on the outskirts of our patch. The appliance had drawn up outside a small cottage and I was checking a water hydrant by placing the equipment on the stand pipe and turning the water on and off again to see if we had water in the mains there.

The rest of the crew and the appliance had moved on to the next hydrant down the road, and I was supposed to walk down to and meet them when I had checked my hydrant in the roadway. I had removed the hydrant lid, attached the standpipe and turned the water on.

Down the garden path of this cottage walked a woman and in a very posh voice said to me, "Excuse me, what are you doing?" You just don't say these things to me with my dry sense of humour. I immediately replied "I am from the water board, madam. I am about to turn off your water supply. Have you filled your bath and kitchen sink with water before I turn it off?" She said that she hadn't, and would I please wait whilst she did so. I said I would.

After some ten minutes, while I pretended to be waiting on her outside the gate, she came back down to thank me and said she had now filled both her bath and the kitchen sink. I said, "That's all right madam, but I thought you would already have done this as it said in the local paper that all supplies to the area would be turned off by three in the afternoon and it is now three fifteen."

She didn't have the local paper that week, which was helpful to my cause, and she asked how long it would be before the water would be turned back on. I replied, "It will likely be tomorrow." I closed down the fire hydrant and departed, leaving her believing that the water supply had been shut off and with no idea at all of who I was. Rotten devil, aren't I?

Chapter Twelve
LOOK WHOSE FLYING

When you are with the same lads every day, and spend a great deal of your working time together, it is not unusual for you all to be involved jointly in some scheme or other.

So it was that some bright lad in the station suggested that, as a team of twelve fire fighters, we could make a bob or two from extra curricular services which we could provide ourselves. This lad's bright idea was that we could buy a bouncy castle and let it out at birthday parties to all the local kids. This way we could make our money back and earn a great big fat profit for ourselves. Nothing more simple than that.

Bouncy castles even in those days were very expensive to buy. But by luck, one lad on the station was reading through the local paper and noticed that a club was selling a bouncy castle second hand and asking only a hundred pounds for it.

That was easy, twelve lads, one hundred pounds between us. We could easily afford that. We contacted the club concerned and found out that according to them, we would need a van of some description because the bouncy castle was quite big. So on the next routine run out to stores, we sent two men from the station to pick up the bouncy castle and put it in the back of the mini bus on their way back.

Roughly three hours later, just after we had arrived back on the station from another fire call, our mini bus turned up. When the two lads had left on their run to the stores, they had been immaculate in dress. Now here they were standing before us, shirts hanging off their backs, soaked in sweat, with damp patches all over the back and front of them, black as the ace of spades, and with hair stuck to their faces.

Naturally, we were wondering what had happened, but that would have to wait until later as we were all eager to see the new purchase we had made for the benefit of all. So we asked them where the bouncy castle was, and without a word from either of them they just pointed to the back of the mini bus.

We ran round to the back of the bus, and opened up the back doors expecting to see a lovely, neatly folded bouncy castle lying there. Instead when we opened the doors, this monster of a thing burst out of the back of the mini bus, a bit at a time, almost killing the first three waiting in line, as it unravelled its way out, piece by piece.

In spite of what we had heard on the phone when we first made enquiries, it would seem that our two friends who went to collect it needed a lot more help than we had ever imagined. And unfortunately we had made a deal with the club to buy the castle 'sold as seen' with no money back.

Somehow the two of them had pushed and pulled it into the mini bus a bit at a time until it was all in the back. Having just about killed themselves in the effort, they then drove off with it. When we got it out of the bus, it took no less than twelve fully fit firemen to lift this thing, such was the weight of it. You can imagine how much laughter abounded that day as it suddenly struck home what we had actually got ourselves into.

Undeterred by this, we set about clearing out the fire station of the motors in there and half dragged, half carried this thing into the station. Now the fire station had four massive bays in it, enough to hold four fire engines, and this thing we had purchased and dragged into the station was now the only thing left in there.

We rigged it up for inflation with two very good condition air pressurised pumps tied into the string ties at the back of the bouncy castle, and plugged it into the electrical power.

As the castle began to inflate, it grew and grew until it had virtually filled the whole of the fire station. But the funniest thing was yet to come. As the floor of the castle began to fill up with air and take shape, it was obvious that it was about three to four feet from the ground and no normal kid was ever going to get into this monster without help.

When the sides of it filled up and began to grow and grow and grow, there were uncontrollable fits of laughter from everyone as the side towers just got bigger and bigger. The station roof stood about twelve feet from the ground to allow the fire appliances in and out of the bays, and here was our bouncy castle still growing on these turrets, which by this time had pushed into the polystyrene tiles of the false ceiling. And with each breath of air that went in, they were still growing.

Remember that the sole purpose of buying this bouncy castle was to let it out and make money from it by setting it up in people's front or back gardens, and here was this monster taking shape in our fire station, still growing, and needing twelve men to lift it. In truth, if you hired it at all, you would need a field to put it in, not a garden.

We weren't going to let a little thing like that put us off though. So a while later after we had stored it in a garage round the back of the station, someone had the idea of using it to make money for the fire brigades charity fund.

This time we set it up with the greatest of difficulty in between buildings in the yard. We got this monster blown up with the machines and I was left in sole charge of it for the first couple of hours, whilst the others were round the other side of the building doing various tasks for a public open day on our station. Initially, we did a roaring trade on the bouncy castle. In fact I'll even go as far as to say that, at twenty pence a go, every kid in the place was on it in the first hour, each being flung on to the front apron by myself, until at one point I had no less than sixty kids on the bouncy castle with still lots of room for more inside.

However, I hadn't realised what the stress of having sixty kids on it at one time would do to the ties at the back of the castle. All of a sudden one of the ties on the blowing machine behind flew off allowing air to escape rapidly. A few moments before, with the aid of their mothers, I had jokingly been throwing their kids gently onto this bouncy castle, and now because the air was escaping rapidly the walls and turrets were collapsing inwards on them.

Being on my own, I could not race round the back and re-attach the blowing machine to the bouncy castle to re-inflate it. So I whipped off my shoes and dived into the middle of this collapsing bouncy castle. Where a few moments ago I had been gently putting the kids onto the apron, I was now very quickly forced to grab an arm and a leg of each in turn and virtually fling them out to the front one by one to screams of fright from both the kids and their mothers. The children came flying out of the front of the bouncy castle for their mothers to grab and some were bowled over as they came.

When the fire fighter whose turn it was next, came to relieve me some two hours later, I looked like the two lads who initially went to collect this thing. Where once I had been the neat and prim fire fighter, I was now completely soaked in sweat. He could not understand it for, as I had now done all of the repair work to get it back into working order, and when he arrived I was simply busy re-attaching the string to the back of the bouncy castle and re-inflating it. He just couldn't work out how I had got in such a state until I told him what had happened.

Sympathy? I got none of it. He just burst out laughing in front of me and ran off to tell the rest of the crew what had happened in their absence. But it taught us that we had to limit the numbers on the bouncy castle to twenty each time and make the rest wait their turn, and from that point we did not have a lot of problems all day long and we made an awful lot of money for our charity. After that day we always had two men on hand in case it happened again.

Eventually there was a lad from another station who wanted to give his kid a birthday party at his home. He did have a field, he'd heard about our bouncy castle and he wanted to hire this thing from us. So all twelve of us lifted it into the back of his pick-up truck, half dragging it in because of the weight, and on that day we had our first, and our last, hire. As it was, he was so pleased by it, that he offered to buy it from us. So we agreed to sell it to him for one hundred pounds, and we got our money back. But what a monster this thing was.

Chapter Thirteen
BEING READY

As you can imagine, because of its very nature, a fire fighter's job is a very demanding one and to be able to do it you have to keep yourself fit. Some calls are protracted and demand a great deal of stamina to be able to endure hours of hard physical work.

Some lads do weight lifting. Some run on their days off. But on station it was usually the job of the officer in charge to find us a suitable physical exercise or two that, although demanding, were not over demanding in case the next call was one that needed all our energy. In the days prior to physical training, with instructors on station, or visiting, we were as often as not to be found clearing the place of appliances and using the space as a volleyball court, getting all the crew on watch to join in.

This worked well until accidents, usually caused by over enthusiasm, set in. One day, the station light covers came crashing down to break on the bay floors after the ball kept continually hitting them. So for a while after that we set up a court in the station yard and played there. But even volleyball can be dangerous in the wrong hands, and someone was always going over on their ankle or being injured by the game and having to go sick the next day as they couldn't walk. Eventually senior officers from headquarters decided that enough was enough and pulled down the curtain on that one.

We were lucky enough to have a very modern sports complex in our area with everything you could wish for. We all became members and for a while every day, we left the driver in the cab in radio contact with our control while the rest of the crew went inside and had a lovely time in the swimming pool. Unfortunately, time and time again, we would have just got into the pool when the driver would come up on the hand held radio we always took in with us and usher us out again for a call out. So eventually that also went by the way.

On one occasion, one of the crew had been reading through books for promotion exams and come across a certain type of boat which could be completely built from materials carried on the motor. So we decided on the next training session to take the motor out to the local nearby river and try and build this boat.

Our appliance held four lengths of hard suction pipes, which are normally used to supply the appliance with water from open sources such as rivers, streams, and dams when no other source of water is available. According to the crew member's instruction, all we needed were a couple of lengths of this big diameter suction which could be linked together, plus a salvage sheet tied round it to cover the middle, and a couple of shovels to act as oars and there we were in our improvised boat.

After we had built this, we decided that the best place to try it out was the deepest part of the river that we could find. To this end we placed the boat in the river and I climbed in with a shovel as a paddle to hold it against the river bank to allow the other two crew to climb in. Because it was this lad's invention, he had to be in the middle of us as Captain. Pulling away from the bank, it seemed to be a wonderful invention and we gaily paddled our way into the deepest part of the river.

Whether or not it was the first time we'd done it, or whether we didn't put it together strongly enough, I'll never know. But when we were over the deepest part, I noticed that our little boat was letting in water at an alarming rate ... and so did the lad at the front. By now the water was up to our waists sitting in this boat, so I decided it was time to abandon ship and promptly did a backward summersault out the back. The other crew member, seeing me do this, did a forward summersault out the front, leaving the inventor of the boat to his fate.

As we were both now in the river, we looked across to the boat to see our wonderful ship's Captain realise that he was stranded out in the middle of the river without any means of steering the boat which was now filled up with water. Like a true Captain, he accepted the inevitable, stood up in the middle of the boat and, with a final salute, promptly sank in the middle of the river holding the salute as the water rose over his head.

If I remember correctly, I believe we spent more time laughing about this than we spent rebuilding the boat on the bank afterwards.

It wasn't always fun like that, of course. I don't know whether it was merely a co-incidence but I had changed brigades from one which had a major hospital for the disadvantaged, as I prefer to call them, to another which had exactly the same thing just up the road.

Inside this hospital, there was an individual who for a period of weeks, all day long, every twenty minutes during the day, whenever he was allowed out of his room, would walk down the corridor of the hospital and without warning put his elbow through the call point. This set off the alarms in the hospital, requiring it to be evacuated on every occasion.

Having turned out from the station every time to check it out before allowing the patients back into the hospital, we eventually narrowed the cause down to this one individual. I won't tell you what we called him but we gave him a name and it stuck with him for many years. At one point, it got so bad that we weren't even making it back to the station before we were being called back to the hospital by another alarm. By now we were all fed up with this so what we eventually did was to hold this bloke against the hospital wall and warn him what we were going to do to him if he didn't stop. No other method seemed to get through to him in his circumstances, but this technique seemed to work, however, as he then moved on to other things.

The hospital had a church linked to it as part of the complex. From time to time, it was also used for funeral services. On one such occasion, the hearse had pulled up outside the church when our little friend (let's call him 'Johnny') was outside taking his usual daily exercise walk. On arrival at the church, the undertaker opened up the back of the hearse and he and the pallbearers carried the coffin up the stairs. They laid the coffin out for the service, remaining inside the church, so that they could remove the body back to the hearse when the service had finished.

A couple of days later we read in the newspaper that our friend had been walking by at the time. He noticed that the back door of this lovely gleaming hearse, which was only three weeks old, had been left open when the undertakers had carried the coffin into the church.

To give him his due, our friend was a very helpful and delightful man who simply wanted to help. So he had shut the back door for them, got into the driver's seat, and without any idea at all of how anything worked, he simply drove off in the hearse.

When the undertakers came out to return the body after the service, both the hearse and our Johnny had disappeared from the front of the church, and there was the hearse three hundred yards further down the hospital complex with its front completely stoved in having crashed straight into the hospital laundry wall. The vehicle was a complete write-off. The undertakers had had to take the coffin back inside the church and hastily arrange for another hearse to pick up the body.

On another occasion there was a fire in the nurses' quarters in the same hospital complex. As I was the driver that evening, I drove up to the nurses' quarters and got the hose reel out of the back locker for the lads who were going in. Then I got the water to be supplied to it from the appliance to fight the fire.

The room concerned was round the other side of the building furthest away from me and out of my view. The rest of the crew had disappeared into the building to assist the lads fighting the fire and I had been left round the back of the appliance watching the radio and gauges as is normal. I was suddenly approached by this smartly dressed gentleman in a clean white coat with a stethoscope round his neck. He asked me pertinent questions about where the fire was and how it had started. He had a very Queen's English type of voice, and being out of sight of the fire and not yet having any radio messages back from the crew at that point, I was trying my best to explain the situation to him.

He was very persistent, however, and continued to ask all the right questions when, at that point, we were approached by a second gentleman in old baggy trousers and an old jumper which looked like it had been thrown on. He came up to the other gentleman and said, "Donald, how many times have I told you not to get into the doctor's uniform? Now get inside the hospital and back into your room".

It turned out that the man was a patient in the hospital and had done this on a regular basis, but you would never have guessed it by the way he spoke that night. The doctor apologised profusely, and we both had a good laugh about it at the time.

In the centre of the hospital complex stood a flagpole in the middle of a grass roundabout. Some of the lads had been there a long time and knew all the characters. On one occasion when I was being shown around, I noticed to one side of the roundabout an old lady with a suitcase. She kept looking at her watch and staring into the distance, while another gentleman at the same time was continually walking round and round the flagpole in the middle. This gentleman would walk round the flagpole and after one complete circuit suddenly turn and bang his head on it, only to repeat the same thing time and time again. So, not knowing what the story was and why these two behaved as did, I just had to ask the crew. When they told me the true story, it made me understand the true sadness of these places.

None of them could tell me why this gentleman was doing the head butting of the flagpole, but they said he did it every day when he was let out for his exercise walk. As far as the old lady was concerned, it would appear that she came from a quite well to do family and her story is truly one of sadness.

In the fifties, when she was young, it would seem that this lady had got involved with a local lad, much to her family's disapproval. In fact they banned her from even seeing him. Secretly though, the two young ones had made arrangements for the young man to come at six o'clock one winter's evening. He would bring a ladder. She in turn would pack a suitcase ready. He would put the ladder against the wall to her bedroom where she could climb down. They would go to the nearest bus stop, and make their way to Gretna Green to be married there without her family's approval. At least, that was the plan.

On the appointed evening, the woman had her suitcase packed, but six o'clock came and went, then seven, with no sign of her young man. In the mistaken belief that she might have missed him, she slipped out of the house with her suitcase, went to the nearest bus stop and stood waiting on her young man. He never showed up.

The stress was so great that she never recovered from her ordeal, which put her over the top and eventually her family had to remove her to this hospital for treatment.

Such was the state of her mind that every day she re-lived the experience and stood outside at the roundabout with her suitcase packed, come rain or shine. In her mind the roundabout must have represented the bus stop and she was literally going back to that fateful evening. Even worse was the realisation that she had been doing this every day since the fifties.

I am told that because the hospital was closing and she would have to go back out into the community, this lady had been taken into a side ward by the doctors and had the situation explained to her. They considered her safe to be released back into the community and asked her where would she like to go? I don't know if her answer shook the doctors or not, but when given the opportunity, this lady simply responded by saying that she had come into the hospital as a young woman and had been there ever since. What did she know of the world outside. All the friends she had ever made were in the hospital with her. She had nowhere to go so she would remain in hospital care until she died.

Sadly I was told that some time later, before the hospital had closed, this lady got her wish and she did die in their care.

Chapter Fourteen
LOOKING AROUND

There were times when you would go out in the fire appliance to drive around and inspect your area. More often than not, when not attending emergencies, you would spend your time looking at buildings where you were unsure of the manufacturing process going on inside, and making appointments with the staff to view exactly what was going on.

Frequently you would also have to visit this type of building to see if all was in order in legislation as per the Fire Precautions Act and inspect their Fire Certificate if the premises had one. One day, I had been allocated a certain bakery to inspect. I had our station officer with me to see how I got on with the inspection.

I had made an appointment in the morning with the manager of the premises, so all I had to do was turn up at the appointed time, very smartly dressed accompanied by my station officer. When we arrived, we were met by a stunning blonde young lady, who said she was there to show us round and handed me the premises' fire certificate. Needless to say, we were both taken aback as she was not our usual representative.

I advised her that all I intended to do was to ensure that the premises complied with what was stated in the fire certificate by looking round. Then I would return to the office where I would inspect the certificate and take down some details from the representative. I inspected the premises and on returning to a side place in the bakery itself, and not the office, I proceeded to ask the young lady for the information I needed to know so that my paperwork could be filled in.

I had intended to start by asking her the address of the premises, but instead, for God alone knows why, the question came out of my mouth as, "Could you let me have your particulars?" And this young lady, being as quick off the mark as I had been throughout my career, immediately came back with, "And would you be wanting my feet size as well?"

The station officer promptly doubled up in fits of laughter and as he tried to control himself, plonked the elbow of his tunic straight into the centre of a very large cream cake they had been making. The young lady, thinking she had got one over on me, was in tears of laughter herself.

I allowed her to calm down and pounced, as she had no idea of what my wit was capable of. Instead, I said I was going to ask her one or two questions to which she should reply honestly, and after calming down she said she was sorry and she would answer honestly. She was obviously expecting to be asked about the premises, and I continued saying I needed to know certain things in order for my paperwork to be filled in correctly. Meanwhile the station officer was cleaning the cream off the elbow of his tunic with a cloth which the young lady had kindly given him.

I started by asking her age, her date of birth, and her star sign. The station officer, seeing what was coming, excused himself from the premises, and went outside only to burst into tears of laughter again while he waited for me. I allowed him to go before continuing with the questions, whilst I pretended very seriously to write the answers on my paperwork in front of the young lady.

She was answering very honestly, believing this was all for the report, while I continued with the rest of the questions. What was her telephone number? No, her personal one? Did she like fast cars? Did she like to go dancing and when was the last time she had been? She replied, "Quite a while," to which I added my final question, "What do you do on Friday's between 8pm and midnight?" "Nothing," she replied.

Having answered all these questions, she was stunned when I immediately came back with, "In that case, that's funny. I have a fast car, so I'll see you at 8pm on Friday," and she couldn't believe how she had been caught herself.

The other kind of premises, of course, is the private one, where you see a house as you drive past, but you don't know what it contains. You may not even know if it is a private house because it's snuggled away behind some very large hedge or fence and you are left to make assumptions.

I had been on station for some years by this time, when we were called to a reported fire alarm ringing at one of these premises. None of us knew exactly where the premises were, although we recognised the name of the street. So off we went at high speed to the lane given in the call out sheet.

On arriving in the lane, we looked around and through a process of elimination, we identified that it had to be this opening in a surrounding bank of earth by the roadside. But there was nothing there, other than this gap. However we took the fire appliance through the gap and were surprised to find a large closed gate in front of us. Surely these had to be the premises we had been called to, or were they? We had to look further.

We opened the gate for the driver and found ourselves on a single dirt track which led to who knows where, but we had to drive up it to find out. We drove up this hedge lines track for about a quarter of a mile when suddenly the track opened out to tarmac. There was a huge car park marked off with lines and there in front of us were the premises we had been called to.

I was in the middle of the cab, which meant I had to accompany the officer in charge of the appliance to investigate the cause of the alarm inside this building. So I duly jumped out of the cab on the side nearest the building while I waited for my sub officer to join me from the other side of the motor.

Whilst I waited, I looked around at this place. It was beautifully laid out with a central green, which must have been all of a hundred yards across. The green had a nice tarmac path around it's border, and there on the far side of the path was the most perfect pear shape of a 'bike stand' you ever saw in your life, big at the top with legs coming down to the ground. She was watering the roses on the border of the path with her back towards us.

I just couldn't believe what I was seeing. She didn't appear to have anything on, which seemed to be the problem. To check that I wasn't seeing things, as soon as my sub joined me I pointed out what I had seen, and he couldn't believe it either. When we both realized what it was, we both burst into laughter, which was when we were approached by this elderly gentleman, wearing nothing but a pair of surgical gloves. He was carrying a pair of scissors in one hand, and in the other was a white rose he had recently cut, whilst his little thing hung freely from his nether regions.

Worse was to come, however. We both struggled hugely to keep a straight face, as he explained that he was the gardener, and how could he help us. He seemingly wasn't aware that the alarm had gone off in the building behind him. When we explained this to him, he said there were people in the building who would allow us to look around and then left us to it.

Luckily, these gentlemen were all foreigners, and having seen us approaching, they had raced off to shove something in the form of shorts on their nether regions. Otherwise we would have been forced to endure the same sight again as the little gardener had shown us and I don't think either of our sides could have withstood this at that point.

Looking around the building, however, we soon established that no alarm was ringing and when we contacted control, they confirmed that a young child had dialled 999 from within the building saying that the alarm was ringing. We established that there was only one child in the building that day and eventually we found him hiding in the rear of the premises. On questioning him, he denied vehemently that he had made the call, but his attitude when he was answering gave him away. Rather than lecture him ourselves, we left him to his parents to deal with and made our way back to the appliance.

As we were turning the motor around, we pointed out the 'bike stand' to the rest of the crew in the cab. She was still watering the roses on the far side of the path and they immediately burst into fits of laughter, not having spotted her before. We had only found ourselves a place that only advertises abroad, whilst being on sovereign soil, and none of us even knew it existed. But we did now.

This had all happened in early spring when things were a bit quiet. So it was that we were travelling up this same lane several months later with a different crew aboard the appliance, including one chap from another station. Only the sub in charge and myself had been on the original callout previously, so I happened to mention what was behind the big hedge there.

Being a disbelieving lot, like they all are, they were having none of this. So I suggested to the sub officer in front, that we should go up the path and allow them to see for themselves, so they could believe me. At first he was having none of this, but after constant heckling from the rest of the crew who still didn't believe us, he reluctantly agreed only to take the appliance up to the closed gate from where we had approached this building on the first occasion. So he ordered the driver to turn into this gap in the embankment.

On reaching the gate the sub was gaily telling the rest of the crew that up beyond this gate was this building we had been to and what it was, when just at that point a car with foreign number plates drew in behind us blocking us from returning out through the gap in the embankment. So there we were, on a single dirt track, pointed towards this closed gate, with this foreign car wishing to get in the gate and us preventing him.

We couldn't go back, so we decided that we would open the gate, travel up the dirt track, which was only just wide enough for the fire appliance, without damaging the bushes on either side, and without drawing too much attention to ourselves. Then we would spin around in the car park at the top, which both the sub and myself knew was there from the previous visit, and get to heck out of it before we really got into trouble for being there in the first place.

That was the plan anyway. What really happened was beyond anyone's imagination in that crew on that day. We were travelling up the track in front of the foreign car, and were about two thirds of the way up when we looked to our right and there was this stunning Scandinavian girl aged about fourteen or fifteen, with beautiful blonde hair, bouncing up and down on a trampoline, while what looked like her younger brother steadied the trampoline as she bounced.

Can you imagine the sight of this beautiful young girl in the altogether, bouncing up and down on the trampoline? And each time that she did so her young breasts were going up with her and in different directions, for she was amply endowed.

By now the crew in the back were in hysterics and the sub officer in front was screaming to shut up or we would be caught. As it was, he didn't have to wait long for that to happen. When we reached the top of the track where it opened out to the tarmac, we saw hundreds of people in front of us, and all in the undress. We'd hit the high season.

That was it. The crew as one were rolling about in the back of the cab in tears of laughter, as was the sub in the front, and the fire truck was going from side to side as the driver couldn't sit upright in his seat from laughing. At that point we were stopped by a man enquiring as to what we were doing there. He had a pair of surgical gloves on his hands, one holding a pair of scissors, and the other, yes you guessed it, holding a rose. It was the very same gardener who had met us the first time, only this time his wee thing was swinging about in the breeze a bit more.

In order to answer the gentleman, the sub officer had to wind down the window in the front of the cab and with the straightest of faces, was explaining to him. He was saying that we had previously been up on a call here, and the crew were wondering if the place had any alternative water supplies other than a hydrant, like a swimming pool, or perhaps a lake at the back of the building.

It was total nonsense, but it was the first excuse that came into the sub officer's head to get us off the hook. I had recognised the old boy from the first time and pointed him out to the crew in the back, telling them all about him as he had approached the motor. Now here we were sitting the back listening to our sub worm his way out of trouble with the biggest load of rubbish we had heard in many years. That was enough for us in the back.

I personally, from the middle of the cab, burst into the biggest bellyaching laugh of all, whilst the sub had his head out of the window, with his most serious face, speaking to the gardener.

My laugh was so loud, that the sub officer while speaking to this man through the window of the cab, with one arm resting on the window, had his other arm inside the cab punching the living daylights out of me. But I couldn't help it. I was on the floor of the cab, biting the sleeve of my fire tunic to try and stop laughing, but it was useless, as I bellyached on the floor. When I looked to one side, I could see a colleague, below the level of the window, on his back on the seat holding his sides, with tears of laughter rolling down his face, yet silent as he was unable to even draw breath because of the laughter. On the other side, another colleague was on the floor with me, face down and rolling about in agony with laughter, so much so that he had fallen into the space for the stairs and couldn't get out for laughing.

How did we ever get away with half of the tricks we got up to?

Chapter Fifteen
A MEANS OF TRANSPORT

When I look back on it, I don't know how I ever survived with half of the means of transport I have used to get back and forward in over two decades.

I was never the lucky one as far as transport was concerned. When I was first employed as a fire fighter with the brigade in Scotland, I lived in a village and my place of work was in the main town a couple of miles away.

In those days it was a matter of catching a bus from the village to the town, which was fine, but the station was on the edge of town and quite some distance from the main bus station. This meant that after getting off the bus, I had to start walking for another ten minutes in all weathers before finally reaching the fire station.

And on a Sunday public transport didn't start running through the village until ten in the morning, which was the exact time I should have started work.

This wasn't exactly convenient. So, having just left the forces after more than twelve years service and having received a final pay out, I had a cheque for over one and a half thousand pounds in my possession. This was enough to buy myself a half decent car from the local garage, albeit that it would still have to be a second hand one.

Eventually, after a lot of trying this one and that, I finally opted to have this Vauxhall saloon, and I reckon that from that day on my 'luck' with automobiles has remained with me.

The salesman, a very nice man, assured me that my choice had been a wise one. Not only was I buying a very reputable car, but if I could wait for three days, he would even throw in a total service of the vehicle before handing it over to me. On the Friday evening, I picked it up from the garage and drove it away from the forecourt trying to accelerate down the road that leads to town.

However, this car, far from being the reputable vehicle the salesman described, just would not open up at all and was holding back something awful when the accelerator was pressed. Because I was eager to take the car home to my first wife and get her and my child aboard to try it out, I put this problem down to dirty petrol, rather than accept that the engine was faulty and return the car to the garage. I didn't have much choice in the matter as when I turned into my own street in the village, only a couple of hundred yards from the house, smoke started to pour from the dashboard in front of me. Not only was the car holding back, but before I could get it home, having only driven it for some two miles, the car was now on fire.

I switched off the engine and the smoke started to disappear from beneath the dash. So I managed to walk the rest of the way home, explained the situation to my wife and called the garage.

My very first car was then towed away and returned to the garage where all sorts of things were done to it, and new bits fitted before it was returned to me a fortnight later.

Whether it was my poor choice, or whether it was the salesman saying privately "Thank God we've got rid of that," I'll never know. But I do know that in all the time I had it, which was about a year, this car continued to hold back engine wise, and it was forever in the garage having this part or that part replaced to try and fix the problem. The garage could never work out what was exactly wrong with it and I was forever the target of the rest of the crew's remarks about the heap that I had bought.

However, one day, when I was returning this vehicle to the garage yet again to have it looked at, it happened that another customer had just sold his caravanette to the garage. Fortunately, through negotiation between the salesman and myself, I was eventually able to get him to agree that if he wanted rid of the problem as much as I did, I would buy the caravanette from him if he gave me the full purchase price of the car as part exchange. The caravanette was a bit more expensive but hopefully much more reliable.

Meanwhile, my wife had got a job in the local university in a village in the opposite direction to my own travel. She also needed transport so I bought her a folding bike which she used initially to get back and forward on. But Scottish winters can be notoriously harsh, and I was having to defrost my wife before getting her to move at all when she returned home.

It got so bad that after a bout of tears one day, I went out and bought her a small motorbike which she could learn to ride. This got her back and forward to work and seemed to do the trick for a couple of weeks. That was until she saw the caravanette I had just brought back from the garage in place of my old heap of a car.

My dream was that every time I was off work, we would simply fill up the water tank with fresh water and my wife and kids would jump in this motor home and we would be off to wherever we wished. And we did wander far and wide in it. However, my wife had also experienced the heater, that lovely heater, and I was relegated to that motorbike which had now suddenly become my means of transport to work.

This was fine in the summer. But one morning I woke up in winter and got ready to go to work. My mother-in-law had warned me that it was snowing heavily outside and I would have to be careful getting to work on the motorbike. Undaunted I set off well wrapped up for the weather. It was only a couple of miles to town, and surely would not take long. Then I would be in the nice warm fire station out of harm's way so to speak.

This was going on in my mind as I set off to work only to discover moments later that the motorbike was sliding all over the place as the wheels would not hold on the treacherous untreated roads beneath me. Very gingerly, very carefully, I made my way to work as the snow built up in front of me, on my clothes, on my face and on my hands, and with the snow blowing directly into my eyes, they were the longest two miles of my life. I promise you that after I pulled into the back yard of the fire station, the remainder of the crew had to come out and assist me off the bike by whatever means possible.

At first, I believed it was simply a matter of switching off the engine and climbing off, but when I went to move I discovered I was frozen solid in the position I had been on the bike all the way to work. I just could not move at all no matter how much I tried.

The rest of the crew, having seen my plight by now, came out to help and between us we worked out that the only means of ever getting off this motorbike was for them to push the bike over and physically drag me off it using my hands to pull me away from the bike. But I couldn't help them as by this time I didn't know where my hands were. I couldn't feel them at all, whilst at the same time I couldn't stop my teeth chattering with the cold.

When they did eventually get me off the bike, it took a further half an hour, and three cups of coffee from the cook upstairs plus virtually sitting on top of the radiator in the station before I could move at all.

Only a short time afterwards, a neighbour was selling his second hand Audi 80, so I traded him this brand new motorbike for it and took possession of his car. What I didn't know at the time that this car was also a heap and within a month of buying it from him

I was replacing first the front wing, then the other one through rust. But at least my wife and I had a car each now and one was a caravanette.

That situation only lasted a couple of months. We were travelling along the motorway in the caravanette on yet another journey for a weekend away, when the engine decided to give out on us. The crankshaft went. I called out the RAC to recover us and we rather unceremoniously got dumped back at our door.

However luck was with us as, some time before, I had bought a spare engine for the motor. I started right away to have the replacement crankshaft from the other engine reground and arranged for a local garage to fit it into place. Only a fortnight later we were taking the caravanette out again after repair, and were only a mile away from the original breakdown, when the oil warning light came on and before I could pull over there was an almighty bang from the engine and we had broken down again.

Once again I called out the RAC and, believe it or not, we had the same man diagnosing the same problem and giving me a lecture on trying to travel on the same engine he had warned me off before. He couldn't believe for a moment that the engine had been completely rebuilt before setting off. He was complaining that it wasn't possible to break down for a second time in such a short period of time unless we had ignored his warning. It was only when he saw for himself the stripped down engine and the missing crankshaft in the garden that he then apologised.

I had the engine skimmed once again and put in new oversized shells this time, as it appeared that the garage in their haste to get the van back to me, had tightened the shell caps but not put the torque to them and one had come off, going up through the engine as it did so, jamming it solidly.

Although repaired once again, the Hi-Ace camper was never the same after that, as the oil warning light on the dash always stayed on. But this was the least of my worries as I had already agreed to trade it in for a very up market camper with seven berths and plenty of head room. The one thing about the Hi-Ace was that it had a fold-up, pull-down head space in the middle of the rear. This was fine but where the cooker was it had a much lower fixed ridge, beneath the built-in cupboards and I was forever banging the top of my head against the steel underside, often coming out of the confrontation with a massive bump and headache afterwards.

Having more or less rolled the Hi-Ace into the garage forecourt and parked it away from the main garage in shame, I handed over the keys to the salesman and took possession of this gleaming seven berth caravanette. Now the sky was the limit and we roamed to virtually every campsite in Britain as far as I can tell. The bed, the kitchen, the headroom and the comfort were all there in this one vehicle.

Sadly, one weekend when we had taken it out, we managed to find the foulest weather possible and this vehicle really did show its true potential that night.

Earlier we had been visiting friends further up North in Scotland and had been watching TV in the caravanette with them when on came a weather forecast about really strong winds expected to gust up to one hundred miles an hour later that evening some time after six o'clock. However, we had plenty of time to get home, or so we thought, so when our friends suggested we go shopping and look around the town, we agreed to go with them. It was well after six by the time we returned to drop them off at their home before departing for our own.

By this time, true to the forecast, the wind was really getting up. But in the town it was not so noticeable, so we set off home by the most direct route. We had only gone a few miles down the road and it had already been a scary journey as the wind caught this motor with its high sides full blast, making it really rock to one side. Suddenly we came to an abrupt stop with police cars halting us. We were told that we couldn't travel the most direct route as several trees had come down on the road we were on and had blocked it completely.

I backed up, got out the map and worked out that if we went down the road by the side of the loch, we could still get home although it was a much longer route. I had already read in the paperwork for this vehicle that it was tested to a tilt angle of sixty degrees. But as we travelled down the loch side road in the open and I was negotiating a bend, a sudden gust of wind well in excess of one hundred miles an hour hit us side on.

One moment all four wheels were firmly on the road and the next the caravanette was completely over at a sixty degree angle, tottering and about to fall over on its side. I could see what was coming as we went up and over onto two wheels on the outside rims of the tyres. At that moment I slammed on the brakes and dived sideways across my wife in the passenger seat. I don't know if it was by diving over my wife, or by slamming on the brakes, or a combination of both, but the side which had been previously up in the air suddenly came crashing back down to earth and we bounced to a halt by the side of the road ashen grey by the whole experience.

After a while we journeyed on with no further ill effects but the wind was so strong by this time that

I couldn't even change to fourth gear as the van was stalling if I tried and we travelled the rest of the journey in third gear. We managed to get to within six miles of our destination before the next gust caught us side on. We were travelling across this not so wide bridge over a river and were in the middle when an almighty bang hit us side on. One moment we were on the road and the next we were physically picked up and deposited on the pavement without touching the ground. Had it not been for the fact the bridge had sides to it, we would have been in the river that night.

That was enough. My wife screamed hysterically with fright, opened the door and immediately got out. So did I, and locked up the van where it was, vowing never to return to it. The next morning, I placed an advert in the local paper and sold the vehicle to two businessmen. No-one should ever have to experience that again, and even now just thinking about it raises the hair on the back of my neck.

Sadly, shortly afterwards, I was to experience the fate of so many firemen and undergo the trauma of divorce and the loss of my children. I would never say that this did not affect me badly because it certainly did. Like so many others before me, I wandered around with a feeling like someone had stuck a knife in the pit of my stomach at the loss of someone I loved so dearly and especially the loss of my children.

So I did not jump straight into bed with the first woman who came along. After a while, I tried seriously to form a relationship with another lady. But having given up the matrimonial home and moved in with her, I knew within a couple of weeks that this was never going to work out and was honest enough to say so to her face. So for five whole years afterwards, I wandered around with this feeling inside me of being stabbed.

That was until one day I accidentally met a woman I had dated as a teenager, twenty years before. We had separated because I was in the army at the time and she was very young. Now she was back and for the first time in a very, very long time I found myself interested again. I have to say that she was still the refreshing girl I had known all those years before, and gradually we got back together. She pulled me back from the brink. Eventually we got back together, and I moved out of my home, leaving behind everything I possessed except twelve black bags of clothes. I was carrying out the one thing I had said to this girl many years before. Now I didn't even have a roof over my head, but once again my new girlfriend came to my rescue, getting a friend to take me in as a lodger to help her out.

To this day that friend remains a friend of us both, and a very good one. Her home was only a few minutes down the road from the station and an easy walk away. It couldn't have been better, and I still had my Audi 80 to fall back on, so I wasn't too badly off for transport either.

However, the curse of the automobile remained with me even then as this rust heap had brown wings, a bottle green body, and was always breaking down on me. So I was always under the bonnet, especially it seemed when I was wearing my best suit.

Having taken it in to work one day, I asked the sub officer if he would like to take a ride home with me as I was going his way when we finished watch. Always eager to accommodate, he jumped in. At the front of the station as you drive out to join the road there is a small dip and my sub officer was a big man. So with him sitting in the passenger seat next to me, we hit this bump hard which was when the Audi decided that something else had to join the two mud wings I had replaced after buying it.

Suddenly without warning, the seat runner, where it joins the seat itself, gave out and as I accelerated up the road sparks were flying from underneath the car and my sub officer was lying next to me at a crazy angle with his feet in the air. He was screaming at me in fright. Serious though it was, I just couldn't stop laughing all the way home until I dropped him off at the door. The car was so badly corroded in the floor pan that it was beyond repair so off it went to the scrap yard which was the only sensible place for it to be.

At this time my girlfriend had gone through a very traumatic divorce herself and had made arrangements to move completely away out of the area. But this time, having said goodbye to her twenty years earlier, and having just re-met her, I wasn't going to let her go again. So I moved in with her and into her life and home with the kids, and between us we purchased a second hand Renault 18. This was a good clean car and very reliable for getting us back and forth to work.

Well that was until one evening we were sitting at the junction of the M8 – M9 Motorway at the roundabout that leads to Edinburgh Airport. I was awaiting the traffic to cross from the right in front of me when some crazy woman looking only to the right and not in front of her, rammed me from the rear at full pelt making a terrible mess of the back of the car and jolting me clean out of my seat.

After the accident we exchanged telephone numbers, but when I tried to contact her later, I could never get her and eventually after months of trying I was forced to use a hire car to get back and forward to work. I then contacted the insurance company whose name she had given me on the day of the accident only to find that she had never informed them of the accident. They said they would contact her instead and from that point on my car was placed in a garage and replaced as good as new.

Having got my lovely Renault returned to me, I vowed never to take the same route again. But only two months later, on my way to work for an evening shift, I found myself sitting at this same roundabout again. Suddenly, right out of nowhere came another car with the driver looking to the right again and not the front as he approached the roundabout. He rammed me as hard as possible up the rear end and once again I was jolted out of my seat.

This time I was furious, having only recently got the car back from the garage, and I was going to give this man the fright of his life. In those days, it was difficult to tell a fire fighter's soft cap from a policeman's because of the badge. Mind you, we looked like bus inspectors too as we all near enough wore the same colours. It wouldn't be the first time when I was standing around waiting for a bus myself, that I had been approached and asked what time the next bus to so and so went.

But I really wanted to give this man the fright of his life, so getting out of the car and putting the cap on I approached him in his car and said, and I quote, "Don't even say a ******* word. You were looking to the right instead of straight ahead as you should have been, weren't you?"

He replied, "How did you know that? I am terribly sorry". I replied "You wouldn't believe it, but I have been ******* well here before and not so long ago. The car has just been returned to me after the same thing on that ******** junction over there. So don't say a word. Just write your name on this piece of paper and be snappy with getting this ******* car into a garage for repair after the last lot."

He wrote his name and address down on the piece of paper as requested. Without a word of a lie, when I looked at it, it read "The Rev. Ian Williamson, The Manse, etc.", and I just didn't know where to stick my own head. I do know that at that point I wanted to eat the cap I had put on to frighten him. But I had said what I had said and there was no going back on that.

Sadly, the insurance underwriter decided that it was one knock too many for the car and wrote it off. So I was left with no transport again. I only got the face value of this car, having sold it on to a friend at a knockdown price. So I was then forced back into the position of borrowing again.

This time, I bought a second hand Avenger which happily took us down to England when we moved South. But that was all. One day, I went to replace the disc brake pads in front of the fire station bay and watched in horror as this car sank to the ground from the axle stands I'd placed it on to hold it up. What had once been the floor pan of the car was now a pile of dust blowing away in the wind.

Dear reader, if at this point you are getting the impression that I've never been one for much luck with cars, you have only heard the half of it. I could tell you about the brand new Lada I bought that broke down on its first journey out, or the oversized panoramic interior mirror that clipped on the original interior mirror, and was so heavy that it always pulled them both off the windscreen when the sun heated it up in the summer. Or the second hand Lada motor I bought as a business car for my worker that was only 1100cc while my own new one was a 1300cc model. Yet the second hand one could lose me in the first hundred yards in my new car. Or how about the brand new, top of the range Lada I bought from under the tarpaulin on its unveiling, with alloy wheels, racing steering wheel, and all mod cons and left it with sister-in-law's man to sell whilst I emigrated. Unfortunately this was the very same time that Lada decided to pull out of the U.K. because of emission problems and this one ended up in the scrap yard unsold.

I don't know how I ever got away with half of them, but just the same I am not going to say it hasn't been a funny experience.

Chapter Sixteen
ILL MET BY MOONLIGHT

I suppose in a way I have been one very lucky person throughout my entire career. I've already described earlier in this book others who moonlighted to earn extra cash, but I have to hold my hand up and say that I have also done it, and successfully, but always with prior approval from the brigades concerned. And with the approval of the taxman as well, for I was always hit hard by him for everything I took on.

However, I have to say that I was doomed to become well known in any area I went into. That is an understatement in itself.

I was once a taxi driver and everybody got to know me so well by sight that they would come out not with, "You're the fire fighter", but "You're our local taxi man."

I was also a professional bingo caller in the new town we had moved to in Scotland and I would get comments about how I was doing the job because I couldn't afford to get married to my new wife. But I was determined to add to the income of a fire fighter, so an extra job was the answer and along came this job of bingo caller, where I would stand up on the stage and call out numbers to packed houses of six hundred people or so.

Even though I had moved away, I was still the one who looked after my mother calling in on her every day. Or on my days off I and my girl, now my wife, would go and get her and bring her over to our home. Then we would all pop down to the same bingo with her. This was her day out and she loved it.

Unfortunately my mum had her leg removed in an operation and was wheelchair bound. She was trapped in the bungalow where she lived, as steps ran down the front and back of the house until I arranged a metal bridge to be built. And this took two years of hassle with the local authority before it was completed.

When I was a bingo caller, the assistant manager was always dreaming up new ideas for pulling in the crowd, and as it was coming up to Mothers' Day again, he decided he would hold a competition called "A Mum in a Million". The idea was that everyone would write an essay on why they thought their mum deserved to be such a person. The prize for this was to be an all expenses paid, long weekend hotel holiday in the North of Scotland in one of this Group's hotels. There would be free bingo thrown in, with a bottle of champagne, a trophy and a bunch of flowers. On top of this, the winning entry would be read out publicly by the assistant manager in front of a packed house.

The entries were all sent to the editorial department of a local newspaper to allow them to judge the winner before returning them to the hall.

By this time, I had made sufficient money to bank and allow us to get married, so about a week before, I had left my job as caller in this hall. My girl kept nudging me and I was eventually coaxed into entering the competition, which I now could as I was no longer an employee.

With a true story to tell and making it heartfelt, I sat down in front of my computer and started to write. My aim wasn't to win but just to tell my mother and others, my own story of this amazing woman who gave up her life to bring up her family of seven children and to tell her on behalf of the whole family how much we all loved her for her choice and the person she was - something she had never had said to her in her life.

I finished the story and, with only one day to spare, I went back to the bingo hall and handed it over to the assistant manager who duly passed it on to the local newspaper editorial staff for judging.

Remembering that the theme was, "A Mum in a Million". I twisted this by starting the piece "Let me introduce you to a mum in a million and her name..........". and I finished with these words, "Let me show you a million and one mum's". On the night of the announcement of the results, we duly fetched Mum from her home and she was seated to the rear of the hall with my girl and myself when the assistant manager took to the stage to read out the winners in reverse order as decided by the newspaper.

First he read out the third place, then the second, and it was at this point I truly did not think we were in with a chance. But my girl kept saying, "You have won. I told you, you would". It wasn't until he said the words, "And the winner is and they even typed out the story nice and neat for me to read", that I knew my mother was about to get her just reward.

I took her up in her wheelchair to receive her prize to the rapturous applause of the audience, and then it was time for the assistant manager to read out the true story I had written. After simply having read out the two previous entries to much well deserved applause, he turned round with mine and announced instead, that the entry had come back from the newspaper people with a note attached. It stated that as they read the story, each in turn was in tears as to its true nature and that it thoroughly deserved to win by far. Then he started to read it out aloud to the audience.

He only got as far as the third line before this grown man abruptly stopped and, choking back tears, apologised to the audience telling them he could not read it aloud, such was the power of its contents that it was bringing him to tears as well. It was never read out that night after all, but those who asked to see it and were handed it to read, also had tears running from their eyes. I had written this essay to Mum from the heart, and I gave it to her to keep.

I'm glad I was given the opportunity then as only months later my mother passed away at the age of seventy eight. She was thirty nine years old when my father died and she did the same length of time again without help in bringing up her family. The trophy she was given that night is with us now and will always be prized as our way of saying how much we all loved her. At least we got the opportunity to say it to her in that essay before she died, so I shall never regret that.

Coming back to other businesses, there was the window cleaning service with the curious name. When we were setting it up, there were initially three partners and we agreed that we should register it with the bank and taxman using the initials of our surnames. One had the initial M, another A, and the third had D, and at first we couldn't agree what order they should take. However, we eventually decided that, as I was taking care of most of the business side, my initial should go first. So it became M.A.D.

We all laughed at this and it was a sure fire winner. You could never forget the name. But we all laughed even more if we tried to turn it around and we imagined what the customer would say as we came up her garden path, had it been turned around. The conversation would have gone from husband to wife, "Hey Jeannie, the DAM window cleaners are here again". I was having nothing to do with that.

The business was totally computerised from day one, right down to invoicing and all the stages before that. It became an instant success, so much so that it wore the two partners out in the first year, and we had to cut the customer list in half as we couldn't cope with the numbers each month. Then the two other partners were bought out by myself leaving for personal reasons and I was soon trading on my own with this business running alongside my fire brigade work.

Unfortunately, I still couldn't cope as the customer list kept rising back to where it was previously. So I employed part time workers, then full time workers and I was still stretched to the limit with this business.

When I first moved down South, the medical care people in the next town were in the process of transferring over to a brand new purpose built hospital where it stands today. I heard on the grapevine that the children's ward were desperately short of toys and books and other things. Even what was there was still in their boxes. Basically they had nothing. So I approached the sister of the children's ward in the new hospital asking if I could get it through the business, what did she need the most?

She gave me a list which was huge. It included sixteen children's videos and she said to me that she realised there was no way I could get all of these things but whatever I could get would be a help. The poor woman never had a clue about how resourceful I could be when I believed wholeheartedly in something. In the very first day I had all sixteen videos wrapped up for her, donated by another business I did the cleaning for. Not only that, but every worker in the place went home and bought a teddy bear or a doll and started to deliver them to my door.

Since this was in mid November, I and my part time helpers stuck on Santa hats and made up certain rules. A customer had given me a plastic moneybox to donate to the hospital, so I capitalised on this by making up the rules like "Whoever you are, wherever you are, if you make comments about our Santa hats, you're copped", and we applied them to the letter.

We were catching the milkman so often on our patch that he took to delivering the milk at three in the morning to avoid us, because it seemed that every time I turned up on a customer's doorstep, the milkman was there and it was costing him an absolute fortune.

Nobody was exempt. At this time I had taken to carrying a big black bin liner to every customer's door explaining what I was collecting and I left the bag on the door in case they wanted to put anything in. Not only were they putting toys in the bag, they were also donating huge sums of money in this little plastic moneybox that I carried with me. This allowed me to buy even more, and soon we had drawing books, coloured pencils, ordinary pencils, and rubbers. You name it, we had it and my garage at home was beginning to fill up with these gifts. But still I wasn't satisfied.

When workmen were digging up leaks in pipe-works near to a customer's home, it was near enough for me to be down that hole with them and they were not getting out until they had put the money into my little box. Tree surgeons were in another customer's back garden and two of them were on my customer's patch. That was enough. The first had money in his pocket and handed it over. The second didn't have any money on him, so I made him borrow from his mate otherwise he wasn't going home that night, of that I was determined. He borrowed and handed it over.

A businessman friend who passed by a customer's door as I was cleaning shouted, "Here, John. I like your Santa hat". He realised he'd made a big mistake as he deposited five pounds for that remark in my little box.

It was at this point that someone leaked what I was doing to the local press and I received a phone call to come to the newspaper's office wearing our Santa hats so we could have a couple of photo's taken and be interviewed by the reporter. She also made the same big mistake as, with all three sitting there, she walked through the door from the office and said, "I like your Santa hats".

So I demanded money from her and after explaining the rules, she popped through the back office, got her purse and dropped the money in the box. But that wasn't enough for me. I wasn't allowed into the back office, but she was. So I sent her with the box to explain that she had jeopardised her fellow workers by her comment and that they all had to pay. Out came all the purses and wallets and in went the money. She just loved the story as I explained what the plan was and every single one of her colleagues either paid up or bought new a teddy bear to put in that sack of mine.

One incident that sticks in my mind concerns the child of another customer. After leaving the bin liner on the doorstep, I suggested that if they wanted to put anything in, while I got on with cleaning windows, that would be fine. When I returned to the door the customer asked, "What should I put in? Can I give money?"

At this point a little girl came out to the door with her mother and in her arms was this big teddy bear. She said to me, and this is the extent and kindness of people's hearts when they reach out to help, " This is Tommy, my favourite bear. I never go to bed without him, ever, but the sick children in the hospital need him more and I would like to give him to them." This little girl was no more than six or seven years old.

One day, as I crossed through a neighbour's back garden to get round to the rear of a customer's house, another little girl was in the next garden playing. On spotting me with a ladder over my shoulder and my Santa hat on, she came running up to me, pulling the sleeve of my jacket. She was four years old at the time, and was saying to me, "Santa, it isn't Christmas Day yet. What are you doing here before then?"

I had to think up something quickly for she really believed in Santa. I said, "Little girl, don't you know? I have come to take back to Lapland the notes of all the children in the world, so they can get the toys and books they want on Christmas Day. I need the ladder to get up to the chimney to collect them. Have you made out your note for me yet"? She said, "No", to which I replied, "If you are quick and get your mummy to help write it out, I will take it with me." So she dashed indoors.

A few minutes later she was back out with her mum with this piece of cardboard. It read "Dear Santa, I want a kitchen, a garage, a rolling pin and a pink elephant". My heart sank for this little girl and on this occasion I will not change her name to honour her. Her name was Amy, and that little girl gave me an insight as to why I was doing this with so much fervour to help others. It was because, like me, she believed in what she was doing. She also told me that her dad had had an accident at work and hurt his leg and her mummy was worried because he couldn't work. So could I use my magic to fix it. With all the will in the world and holding back tears myself I said to her, "I will use all the magic I have to try and fix your daddy's leg very soon so that you all can have a lovely Christmas." What else could I say to such a wonderful little girl.

I honestly could not think of a better reason or a better person than this little girl to hand over all the stuff I was accumulating for the children's ward. I immediately approached her mother saying that because she believed in Santa so much, I would like to make her Christmas dream come true. I would be delighted on behalf of all my customers who had given so freely if she were the person to hand over all these goods to the hospital. I also, asked if it were possible to give her all the toys on the list out of my own pocket but to do it in a very special way.

I wasn't finished yet. The station I worked at had a complete Santa outfit we used for our own Christmas party for our own children. So I roped in the assistance of a part time lad to dress up as Santa, arrive at the hospital and go around the children's ward handing over an electronic game I had bought myself to every child who had to remain in hospital over Christmas. But when he did so, he had to arrive from the other end of the hospital, out of sight of our little Amy and not until she had handed over all the gifts we had collected in these sacks. These had by now reached twelve full sacks of toys and books and games for the children, enough to stock the ward ten times over. We also had all the children's videos the sister had asked for.

On the morning we had arranged, the ward sister was off duty but had arranged for the hospital administrator to take her place. So we duly arrived with these twelve sacks on the back of a fire brigade lorry which I had commandeered with the help of another colleague of mine from the station. Then we loaded them into a side room of the ward because of hygiene considerations in the main ward.

With the help of her mum and dad, who had come along to record the moment on video themselves, Amy then presented the hospital administrator with all the toys. She handed over the sacks which lay everywhere and gave a little speech which she had made up herself with no prompting. Although these were the words of a four year old, I couldn't have put them better myself. She said, "Nurse, we would like to hand over all these toys to the sick children in this hospital on behalf of all the kind people who gave them and we hope that all the sick children will get better really soon." Then she also handed over the little plastic moneybox that started it all with the plastic key that opened it.

The administrator, not realising it had such significance, any more than the toys, took the moneybox and placed it on the bed alongside the rest of the sacks. So I asked Amy to tell her to open the box with the key, which she did and inside was a cheque from myself for every penny I had earned from the first day in tips from my customers. The cheque was for a further two hundred and fifty pounds. I don't know if the lady administrator was completely flummoxed or lost for words but she simply reached down, picked up little Amy in her arms and gave her the most heartfelt kiss on the cheek for all her bravery.

But still we weren't finished just yet. I had primed little Amy's mother to have her hang about the ward door after the presentation so she would spot my Santa come in. As sure as fate, my bright little button spotted him right away and firmly ran straight up to him pulling at his sleeve and screaming, "Santa I'm here". I had also primed my Santa to say he was here for the other children in hospital really but would she like to go round with him and give the children their present of the electronic game for each of them.

We had all joined him by now, and Amy was doing her bit again, handing over the present to each child and saying, "Hurry up and get better soon so you can go out to play." I had brought with me a Polaroid instant camera and took a picture of each as Amy handed over the gift with Santa, never letting go of his sleeve, and that day we left every single patient with a memory of the gift in a photo.

The final twist was that I had primed my Santa that when it was over, to begin to leave by walking down the corridor of the hospital. But my little Amy was having none of this and was pulling firmly and insistently all the time saying, "Santa, what about me?"

I had previously instructed my Santa to say, "Let go little girl, I have lots more children to visit before Christmas Day and I must get on," and my friend played the part beautifully right down to the last word. As little Amy let go of his sleeve with tears in her eyes, he started to walk down the corridor. It was at that point I allowed him to go a little distance before shouting to him, Santa, you still have something left in your sack." and he placed his sack down on the corridor floor and looked inside saying, "So I have.

He brought out the rolling pin and said, "It says 'Amy' here, but I don't know of any little girl here in the hospital by that name. It must be a mistake." The lady hospital administrator latched on to what was going on right away and guided us into the side ward. Throughout all this Amy was excitedly shouting at

Santa "That's me! That's me!" and Santa had replied "Well, I'm sorry little girl. I almost missed you."

We all went into the side ward as gift after gift came out of the sack for Amy, everything she had written on that piece of cardboard paper in the back garden, right down to that God forsaken pink elephant she asked for. I had had all sorts of problems finding one, but nothing is too much when you believe so much. One very believing, very deserving little girl received everything she'd asked for that Christmas. Without doubt, you could tell from the little sparkling eyes and the beaming smile on her face that her memory of that day would last a lifetime and she would to hang on to her belief in the magic of Christmas for many more years to come.

Nevertheless I frequently found myself in all sorts of trouble through this window cleaning business. I was never the one to stop at the hometown for business and before long covered my customer base covered virtually the whole of the South East of England.

In addition, I had bought a luxury static home on a park in Wales a couple of years before, and when I was not using it myself, had been letting this out complete with its own uniformed hostess and cleaner to greet you when you arrived and wish you farewell when you went home after the holiday. This was so successful that it reached the point where I would simply reserve myself and my wife a couple of weeks in the summer and let it out the rest to customers, having it looked after by my hostess. She lived locally and would run along the road whenever the static home was let. It wasn't long before word got around of how this static home had the most fantastic views on the park. It looked right out to the bay and island in front, and was situated overlooking the sandy beach. And then my troubles started.

Business was so brisk that even when people were trying to hire other static homes in this holiday park, the girls in the office were saying, "Look, yes we have a static home to let. But there is one very beautiful home right on the water's edge that is unsurpassed, with its own hostess. So before you hire from us, give this man a call from the telephone kiosk outside and see if his is available. You will not regret it".

I was inundated and before long my wife and I found ourselves giving up our holiday slots and having to take our own holidays out of season.

Around this time, I put to the manager of this park that I was a professional window cleaner with all the necessary gear, know how and ability to turn his site into the most beautifully kept park in Wales. I proposed monthly cleaning of each owner's home, simply invoicing each owner in the office afterwards.

They would then send their cheque to me once a month after each clean.

The manager was hooked and so we started. The idea was that on my four days off each month, I would travel to Pembrokeshire and clean each home on the park from end to end, whilst my own staff kept the business running in the South East. It went well for the first couple of months in the summer, but my static home on the park was now so fully booked that I could no longer use it. However, undaunted, I bought myself a little one man tent and pitched it in a camping site opposite the park.

Instantly my service made a tremendous difference to business in the park. These beautiful clean windows sparkled in the sunshine and made the place stand out from everything around. September came and I was still using this one man tent with a cot bed and a sleeping bag which touched the inside of the tent. With the morning dews, the condensation on the inside was now beginning to seep into the sleeping bag and my feet were always soaking wet. I bumped into one of the few who had braved such weather conditions on this park before. However, they were in a much better situation than I with their raised little trailer tent keeping the whole family well off the ground. They were also in a much larger area than I was. It wasn't long before we struck up a conversation and they asked me what I was doing here in the park. In return I asked them about their holiday.

This couple were called Jan and John. They had two boys with them and this was their second week in the park, their one and only annual holiday together. They came from the pottery towns in Staffordshire and didn't have much money left on their second week so they took to playing cricket on the park to keep the boys amused, and I was roped into to help.

Then I was invited to join them to play cards in their tent on the first evening and since, unlike them, I wasn't short of funds, I suggested that I pop over the road to my own caravan park shop and purchase some sparkling wine and ingredients for a barbeque. They were happy to cook these on their barbeque for the boys to keep them occupied and although they didn't have much, the barbeque was a tremendous success and the boys waded in.

The next morning I was awakened with a cup of hot coffee by John who had brought it over to my little one man tent. I can tell you that since my feet were yet again soaking on the end of my sleeping bag, it was more than welcome.

Each morning after this I received a coffee until their funds ran out on the third morning and they had no coffee left to give. I only discovered this when their boy blurted it out, much to his mother's dismay. They had given me their last without even asking for a "Thank you".

I had really got to know them well by this time. I had seen John looking sad because the money was running low. I had seen Jan retire into her tent in tears because she couldn't give the boys the holiday she would have liked. So I was determined to reverse this situation for this wonderful couple. I told Jan that, when I had finished cleaning windows in the park opposite that very evening, she must have herself and her family spruced up to the very highest standard. She and her family were about to experience a night out they had never known in their lives. In her own words, all she had known was this trailer tent and the fact that she would love to go to a park for a holiday with her boys. But in truth she also knew that she couldn't really afford it.

I couldn't let that go unnoticed and I laid my plans.

When I had finished for the day, I returned to the park opposite and met up with this very loving and warm family. I was stunned by in the fine regalia they were wearing when they greeted me. After I had had a shower, we all made our way over to this holiday park opposite. During the day I had phoned my receptionist down the road from the park to find out from her whether my home had been let out.

At this point, I must say that when the subject of my luxury static home in the park opposite had come up previously, it was met by some scepticism and, I believe, dismissed by the couple. I mean, why would a man be living in a one man tent when there was luxury next door with a hot and cold shower and a comfortable bed?

However, their view began to change as we crossed the road with the boys and the gateman on the gate said to me, "Hello sir. How are you this evening?" Of course, I knew this man well, and his security dog bounded up to me as soon as it saw me, jumping up on me as it always did playfully.

Jan turned round to me and said, "How does he know you? It must be because you are doing the windows." But you could see the look of doubt creep into her face.

I was taking them to the owners' club without telling them where they were going and I still had the special key in my pocket. But on the way I decided to take a detour down to my static home on the beach front. On arrival I explained that this was my static home, but I could not get in it as it was currently let out for the week.

Both of them looked at this beautiful home with its stained wooden veranda, all forty two feet of it, reaching out over the slope to the beach. Then they burst out laughing saying, "Never in a million years." So I asked what they thought of it. They simply would not believe me, no matter what I said. That is until I reached into my pocket and pulled out the piece of paper I had written earlier that day.

Placing it into Jan's hand I said, "Jan and John, without asking, you have taken me into your family, had me join in everything. From the goodness of your heart, you have given me your last, and not expected anything in return. Now it is my turn. You see, I do not work like that. Jan, the holiday you always wanted, but could never afford for your children, is yours for one week from the seventeenth of October. That's the first available date this home is free of hire. Please call my receptionist when you intend to arrive from Staffordshire on that date."

They both looked at this home, absolutely stunned, no longer disbelieving, but not wanting to believe in case it were not true. But Jan suddenly burst into tears again at the prospect. She was a very honest and sincere woman, who cared everything for her family and here she was crying, "Oh, my God, my mascara", as it ran down her face.

I was expecting no more than a handshake from John. But what I actually got was no less than a bear hug from him with tears in his eyes, and no words.

It didn't end there. A few minutes later we reached the owners' club where I was taking them that evening. I opened the door to the most fantastic room, with carpets well over your ankles, leather suites and an open fireplace with a blazing fire to sit around. The club had its own bar with barman, a gymnasium, jacuzzi, steam room, and sauna. Next door entertainment was going on in the holidaymaker's area, and for the very first time I saw a look on their faces as if to say "Oh my God, what have we done to deserve this?" And I thought if you don't know now Jan and John, just go through life being the same people you are now, and maybe someone else will one day come along when you least expect it.

Jan and John took their holiday on the date provided, and brought Jan's mum and dad with them. I had bought all the essentials, like milk and coffee, and half a dozen bottles of sparkling wine for them to enjoy on their holiday. I had also provided a cricket bat and ball, and a football for the boys to use on the beach and had my receptionist put them in the home before their arrival. I left them a note explaining that this was my thank you for their kindness to me. In return they sent us a "Thank You" letter from them all with photos of them enjoying themselves. Even then Jan did not know exactly what to write because she was still stunned by it all.

It wasn't the last time I was to use this home to provide hope. I discovered that the son of one of my own work mates had cancer at a very young age, and for months had to console him, as they struggled from hospital visit to hospital visit to get his cancer into remission. I used a slot in the hiring to send this family away from the stress so they could be a family again in my holiday home. When they returned home a week later, they were a different family to the one that left.

Yes, I have had it all, but I have used it wisely, understanding the people I have helped because of it.

Some time later my wife decided to join me as I cleaned the van homes in this same park. By this time, we had moved up to a four man tent that I had borrowed from my son-in-law, with camping cooker, table, chairs, all to make life easy.

My wife's first experience of camping was also her last. The day we arrived to set up the tent it was pouring of rain, and it never stopped. We were soaked to the skin as we struggled to get the tent up. It was so bad that the park had turned to ankle deep mud and, unfortunately, the toilets and shower room were over the other side of the park away from the tented area.

To get to the toilet in the middle of the night meant getting up with my wife and walking her over and back again in the pouring rain and through the mud. At least we had a blow up air bed which made this tent much more comfortable, but, other than for a couple of hours out of four days, it never stopped pouring of rain. And we were miserable in that little tent.

On the evening of the third day, my wife had been sitting on the beach on the other park beneath our home and the rain had gone off for a couple of hours. She said to me that our mobile home did not look like it was occupied, although it should have been let out for the week. Later that evening we went to the owners' club just to cheer ourselves up and when the security guard came in, I asked him if it was occupied.

He explained that it should have been, but because the couple that had booked had illness at the last moment, they had cancelled and it was empty. He then said did we want to use our home, and would we like him to go to the office and get the spare set of keys?

Did we ever want to use it? After three lousy days soaking wet, did we want to swap our tent for a beautiful home on the beach, with a hot shower and a lovely soft bed to sleep in? You bet we did!

I don't think I even finished my drink as I raced across the other side of the road to strike the tent down in the other park. In complete darkness and with mud everywhere, I started to pack this tent away into the car boot. It was a dreadful job. Once again the torrential rain had come on again and was pouring down as I struggled to pull the tent down on my own. By this time my wife was having nothing to do with it and had gone straight to our static home.

When I got to dismantling the main tent itself, I realised how bad it had been for, as I lifted the main tent with its sewn in groundsheet to pack it away, a toad and a millipede had both moved in underneath to shelter with us. Both scampered as I lifted the groundsheet as if to say, "What do we do now in this utter flood?"

EPILOGUE

At the end of this book, I can truly say that I know I have had a very worthwhile career.

However, after two decades of the privilege of serving as a fire fighter and getting up to the most phenomenal tricks, the reader must surely appreciate that, having divulged the most hilarious encounters, I could also fill a further book with the horror stories of the fire brigade.

I have deliberately steered clear of these stories to make this book one of the funniest true tales around and one fire fighter's personal reminiscences.

I have not only had the opportunity to save life, but also the privilege of helping others less fortunate than myself. And the pleasure of being Santa to the children of the men at the fire station, not once, but several times, arriving on the back of a special appliance for them. Nothing, and I defy anyone to argue with me, nothing gives you a more rewarding feeling than to see the look of happiness on the face of a small child, or the gratitude for a kindness that has been repaid.

Except, perhaps, that one lone person you have saved from death, who comes back after many months in hospital to say, "Thank you for saving my life." I was that lucky. Now I have moved on and the younger lads have taken over, and at the end of their careers they will each have their own personal stories to tell. I hope they share them with you.

So, at the end of the day, has it been worth it, to give up so much of one's life for? Yes, every single second of it.

THE END (Never).